The Jefferson Nickel Analyst

A Comprehensive Analysis of the Series

By
Bernard Nagengast

Published by E&T Kointainer Company

E&T Kointainer Co.
PO Box 103 Sidney OH 45365
937-394-4812

ISBN 0-9719899-0-7

Contents

Credits

There are many who encouraged me to write these editions of The Jefferson Nickel Analyst.

My wife Carol has always supported me, and she is the one who has kept me moving on this project.

Respected numismatists Tom Delorey and Bill Fivaz have provided assistance with many of the illustrations. Marilyn (Tiernan) Van Allen gave considerable assistance in setting up the first edition. The late Walter Breen and Adolf Weiss had always supported and encouraged my quest for more information about Jefferson Nickels. Renowned dealer Harry Forman taught me a lot about the heyday of BU roll collecting. (He sold quite a few to me too!)

Although they did not directly contribute to this work, I must thank the many published numismatic scholars, both deceased and living, whose work impressed on me the importance of spreading numismatic knowledge.

And I want to thank all of you who are reading this book. Let this be only a start to expanding your numismatic knowledge and pleasure!

Author Information

Bernard Nagengast has actively collected coins for almost forty years. He began collecting Jefferson Nickels in the early 1970's, with a goal of assembling an uncirculated set with the highest quality possible. That desire led to a search through hundreds of thousands of uncirculated nickels of all dates and mints. Enough of that search had been completed by 1978, that he decided to publish what he had learned about the series. The result was *The Jefferson Nickel Analyst*, published in three parts beginning in 1979. Later, he contributed to Walter Breen's Encyclopedia of US and Colonial Coins. Nagengast has also extensively studied and written on the art and science of numismatic storage, and is owner of E&T Kointainer Company, a provider of archival quality holders and accessories. He is a member of The American Numismatic Association, the Central States Numismatic Society, and the Shelby County Coin Club.

Outside of numismatics, Nagengast is an environmental engineer, specializing in the history of heating, ventilating, air conditioning and refrigeration. He has published a book and several papers on the subject and was a contributor to the recently published Macmillan Encyclopedia of Energy. He recently appeared in *Modern Marvels: Fire and Ice* on The History Channel. In the past, he authored an economics pamphlet *Hyperinflation: How You Can Come Out Ahead*, issued by Forecaster Publishing Co.

Introduction to The Second Edition

Twenty-three years have passed since the first edition of the Jefferson Nickel Analyst. During that time, the Jefferson Nickel series has undergone some more minor design modifications, but the series continues on (much to my amazement). Several major events have taken place however.

The most monumental one is that full step nickels have been recognized by the major grading services. The prediction some years back was: "That will never happen." Now full step nickel collecting is officially recognized alongside full split band dimes and full head quarters.

Another change is that the earliest advocate of full step nickel collecting, Adolf Weiss' PAK Full Step Nickel Club, has long disbanded. It was succeeded by CONECA's Jefferson Nickel "club within a club" operated by Rich and Sue Sisti. At the same time, Michael Westcott formed The American Nickel Collectors Association that brought together collectors of all series of five cent nickel coins, including Jeffersons. That club disbanded after a couple years, and for a time there was no nickel organization. Recently Darrell Crane founded The Jefferson Full Step Nickel Club, and as of this writing, the new club is steadily growing in size and purpose.

The last major event is that price guides have begun to list Jefferson Nickels in full step grade. Currently both *Numismatic News* and *Coin World* list prices for MS 65 five step nickels, although the prices are still in a state of flux. Can the *Coin Dealer Newsletter* and "The Red Book" be far behind?

Many collectors have repeatedly asked me to update The Jefferson Nickel Analyst, and so I have done so, revising old and adding new information.

I present this to you in hopes that it will provide a comprehensive guide for a collector new to Jefferson Nickels, but also will add to the already extensive knowledge of an advanced collector.

Bernard Nagengast
January 2002

General Information

History

A competition for a new nickel design was held in early 1938 resulting in 390 entries. The winning design by Felix Schlag was announced in April 1938. Schlag was then required by the Commission on Fine Arts to redesign his 3/4 view of Monticello on the reverse to a full face view and to make some modifications in the obverse design. He was also asked to change the lettering style. The redesigned reverse bears a striking resemblance to reverse designs submitted by Marcello Rotundo and Franz Karl Hejda during the contest. Production of the new design began in September 1938.

The original design for the Jefferson Nickel

In 1939, a decision was made to modify the reverse slightly, resulting in more sharply defined steps on Monticello. Several other changes in the design were made in subsequent years. These are covered in the section on design changes.

The composition of Jefferson Nickels was changed in 1942 from copper-nickel to a "billion" alloy of 56% copper, 35% silver and 9% manganese. This alloy was arrived at after some experimentation by the mint, commencing in the Spring of 1941. These experiments included the striking of sample coins of various silver alloys, including one of 50% silver. None of these experimental coins is known to exist.

The "wartime" silver alloy nickels were struck through the 1945 date. The composition then reverted to the copper-nickel alloy previously used.

History of Full Step Nickel Collecting

Collectors of Jefferson Nickels apparently began to notice that some coins lacked proper details, causing them to search for specimens as close to perfection as possible. This probably began in the 1940's, and there are some collectors who said that they were doing this at that time. However there was no formal Jefferson Nickel collecting fraternity until 1977.

The PAK Full Step Nickel Club

Three early collectors of full strike nickels decided to form a club for that purpose in 1977. Philip Petrillo, Adolf Weiss and Karl Nenninger organized the PAK Full Step Nickel Club in Montclair, New Jersey, using the initials of their first names in the acronym "PAK." Nenninger dropped out of the club within a year, but Petrillo continued for several years. Weiss and Petrillo were part time coin dealers. Weiss actually operated the club, assuming total control by about 1980. There were no club officers as Weiss performed all duties. As the club grew in size, he operated with the assistance of Jay Ess, and later, Rich and Sue Sisti. By 1986, there were approximately 1000 club members. The clubs accomplishments included a monthly newsletter and a young numismatist essay contest featuring as a prize, a paid trip to the American Numismatic Association Summer Seminar.

Mr. Weiss was operated on for eye cataracts in 1986 and soon after he left Montclair, taking all club materials with him and leaving no forwarding address. The reasons for his disappearance are a mystery. Some years later he was seen at some coin shows in California, however he never explained why he abandoned the club. He passed away in 1989 in the Portland Oregon area.

The Full Step Jefferson Nickel Club

Richard and Sue Sisti, who had assisted Adolf Weiss in the latter years of the PAK Club, attempted to carry on, later reorganizing the club within the CONECA club in late 1989. CONECA, a national numismatic club focusing on error coins, had a unique structure that included a "club within a club" feature for groups who enjoyed a sub-specialty in numismatics. The Full Step Jefferson Nickel Club began publishing a quarterly newsletter under the CONECA umbrella. The club held periodic auctions, and produced a 35 minute full step nickel grading video. The club continued into the 1990's.

The Full Straps Roosevelt Dime Club

Maryland collector Steven Ritter formed The Full Straps Roosevelt Dime Club in 1981 as a specialty club for collectors of Roosevelt dimes and full step Jefferson nickels. Ritter published a newsletter and conducted auction sales.
The club disbanded at the end of 1982, at that time having approximately 150 members.

The American Nickel Collectors Association

About 1987, Michael Westcott, a 15 year old student, formed The American Nickel Collectors Association out of his home in South Carolina. The club catered to the interests of all nickel five cent collectors, including Jefferson specialists, who comprised about half of the club's membership of several hundred. The club issued a periodic newsletter, *Nickel News*, and Westcott authored a book, *The United States Nickel Five-Cent Piece*, published by Bowers & Merena in 1991. The American Nickel Collectors Association has ceased operation.

The Full Step Nickel Club

Collector Darryl Crane formed The Full Step Nickel Club in September 2000 in Burbank California. The club publishes a quarterly newsletter, and as of this writing is growing rapidly. The club listed their functions as:
1. Creation of a pricing structure for full-step Jefferson Nickels.
2. A standardized grading system for full-step nickels.
3. Research of error and variety nickels featuring articles written by well-known names in the error and variety field.
4. Scarcity and rarity report of full-step Jefferson Nickels.
5. Future pricing guide
6. FSNC auction held two or three times yearly where the highest quality full step nickels are offered to members.

Currently, club dues are $20.00 per year. The club may be contacted at PO Box 10909, Burbank CA 91510-0909.

CHARACTERISTICS

The Jefferson Nickel presents a challenge to the numismatist due to its wide range of appearance, which can vary from eye-popping to downright horrible. Not only does the depth of the "strike" vary, but also the surface smoothness and color.

Copper-nickel coins tax minting technology to its limits. The alloy is hard, and it becomes even harder during the process of punching out planchets from rolled copper-nickel strip. If not properly annealed (softened) the metal will not flow evenly under striking, and die life will be severely shortened. As with earlier series of copper-nickel coinage, the Jefferson Nickel was plagued with minting problems through its history.

Striking Problems

The Jefferson Nickel design suffers from an inherent flaw. The highest point on the obverse, Jefferson's hair above his ear, was set exactly opposite the step area on the reverse. Unless die pressure was high, the distance between dies at the lowest point of travel and/or the planchets were soft enough; the steps and/or the hair at the ear would not be fully struck up.

Softness of the planchets was controlled by a process called annealing. The mint anneals planchets by passing them through a controlled atmosphere oven that heats the metal to a set temperature and then allows the planchets to cool at a controlled rate. If there is any variation in atmosphere, temperature and the rate of heating and cooling, the hardness of the planchet can be affected, and ultimately the degree to which the design elements of the coin are brought up under die pressure.

One would expect that a soft planchet struck with proper die pressure and spacing would result in a perfect coin. That was not always the case. If the coin die was weakly impressed to begin with, the result would be an entire production run of "weak strikes" from that die.

Even a new working die can have an inherent weakness in design details that is dependent on the care with which it is made as well as the condition of the hub from which it is impressed. During most of the time span of Jefferson Nickel production, the standard practice was to use a master hub, produced from the original plaster designs, to impress master dies. The master dies were then used to impress working hubs, which were then used to impress the working dies actually used to strike the coins.

Depending on the amount of use and preparation, any of the dies or hubs used in the sequence from master hub to working die can become worn, and that wear is then transferred to every die or hub made subsequently. No doubt this happened at times, and in fact, the master hub produced in 1939 was so badly worn by the late 1960's that it had to be replaced! Thus there are differences in the strike quality of Jefferson Nickels that seem to vary considerably within any given year or mint.

Quality of the copper-nickel strip used for planchets and well as the quality of the steel used to make the coinage dies has an effect on the ultimate product. Both strip and die steel quality are known to have varied over the years, particularly in the 1950's.

I have often found heavy amounts of gray to black grease, containing metal dust, on uncirculated nickels. Almost always this grease is found with dates and mints that show lack of design details. The grease probably resulted from a gradual buildup of oil, lint and planchet and die metal on the dies that was transferred to the coin. It is known that the coinage dies were wiped with oily rags It has been reported that graphite was used on dies to lengthen their life, and this material no doubt would be part of the grease mix. One can thus picture how debris could build up in the dies. When this greasy material lodges in the deepest recesses of the coining dies, planchet metal cannot flow into those design elements. The result is the lack of detail in obverse hair, and reverse details on Monticello, particularly the porch steps, pillars and triangle. Filled dies are common on Jefferson Nickels from the 1940's through the 1960's.

A stunning example of a nickel struck from a grease-filled die

Worn Dies

The coinage dies wear under use, and this affects the sharpness of the coins produced. The working dies that actually produce the coins are changed out after wear is evident. Of course how often they are changed varies with mint policies in effect at any given time, as well as the attentiveness of the coining press operator. These vagaries combined to give the Jefferson collector a wide range of coin quality from year to year and from mint to mint.

Die wear appears in different years and mints in a non-uniform way that is not consistent. The policies concerning die life and attempted methods of lengthening die life through polishing, etc. were varied. Die preparation methods, type and quality of die steel changed too.

The result is that Jefferson Nickels struck from worn dies exhibit many different appearances depending on year and mint. Early San Francisco mint coins often show field wear or mushy details. Many Denver and Philadelphia Mint coins from the 1940's and 1950's have mushy details. The coins of the 1960's exhibit worn lettering and die-erosion lines around Jefferson and Monticello and around the edge. Many coins of the 1980's have a severe "orange-peel" effect from severe die erosion that resulted when the coinage dies were improperly heat treated.

The following series of photos show the difference between two 1960 nickels, one from relatively new dies and the other struck from a worn die. Notice the erosion lines around the periphery of the coin.

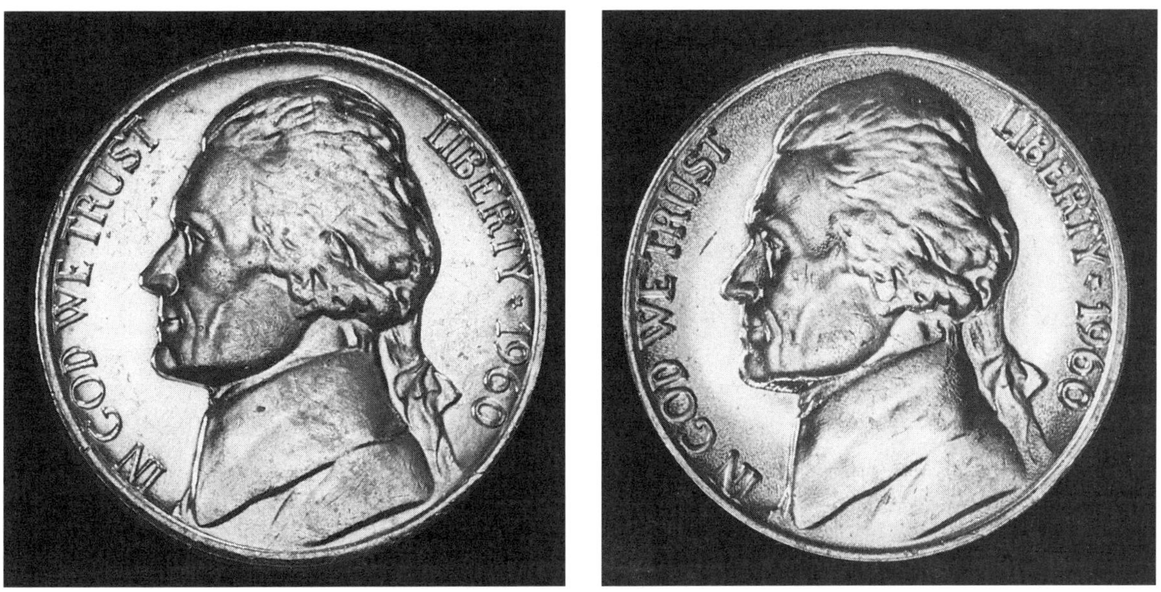

1960 obverse: new die on left, worn die on right

1960 reverse: new die on left, worn die on right

Dark Color

Most Jefferson Nickels are brilliant, however some coins minted by the Philadelphia Mint in the 1950's are very dark in color, ranging from gray to charcoal black.

The Philadelphia Mint in the 1950's was known to be having annealing oven problems, particularly with the atmosphere in the ovens. The annealing process normally causes planchets to darken, and for this reason they may be washed and tumbled dry to brighten them. It is possible that at times, planchets were much darker than usual, and/or the post-annealing washing/tumbling was done haphazardly or even not done at all. The fact is that many of the coins produced at Philadelphia in the 1950's are dark and quite ugly in appearance.

Dark Philadelphia coins seem to have sporadically appeared after 1945, as I have seen examples from 1946, 1948 and 1949. However dark coins from 1946 through 1950 are uncommon. The situation began to change in 1951, when many more dark nickels appear to have been released. After 1951, the situation got generally worse. Much of the 1952, 1953 and 1954 issues are dark. In 1955 things sunk to a new low, and virtually all of the 1955 Philadelphia coins have a darkish cast. For some reason, 1956 saw a sudden improvement, since most '56 nickels are brilliant. Not satisfied with success, the Philadelphia mint returned to whatever it had been doing before, resulting in steady deterioration of coin quality in 1957 and 1958. In 1959, bright white coins became more prevalent. After that, the mint apparently solved its dark coin problem, for I have never seen any dark nickels after 1959.

Curiously, the Philadelphia Mint did such a uniformly good job in producing black nickels in 1958, that some enterprising coin dealers promoted the coins, selling them as "Black Beauties." These coins actually have a certain eye-appeal, being uniformly glossy or satiny black. There are some 1959 black beauties as well, but they are much less common.

Sometimes Jefferson Nickels have a band of darkness across the surface, the area outside the band being brilliant. I suspect that counting or roll wrapping machines with rubber or oil-coated steel rollers came in contact with the coin, leaving a residue that darkened the coin at that point.

Gray, "Circulated" Appearance

Many San Francisco nickels exhibit a grayish color combined with a unique appearance of being slightly circulated. The effect is seen on S mint issues from 1938 through 1941 and from 1946 through 1952. It is so prevalent on 1940 and 1941 issues, that some dealers were convinced that some unscrupulous person had assembled "slider rolls" of AU coins. The effect is also very common on 1947 S nickels as well. I am convinced that the so-called sliders are in fact uncirculated. Close inspection shows no wear at all, and I have seen varying numbers of these coins in original BU rolls.

The cause is a mystery. The possibilities are poor metal quality, and unusual die wear.

Grainy, Rough Appearance

A grainy appearance is common to Philadelphia minted coins from 1941 through 1950, especially on the copper nickel issues. There are several possibilities for this. The planchets themselves may have been grainy and rough resulting from coinage strip rollers that were not smooth. Another possibility is that the copper-nickel itself was of poor quality, perhaps from the inclusion of scrap metal during and after World War II. This metal may have not rolled out smooth, or may have generated lots of chaff that stuck to the strip rollers, which imparted the roughness to the rolled strip.

Poor metal quality could have resulted in dirty coining dies. If the planchets generated a lot of chaff that stuck to oily dies, the resulting roughness would have been imparted to the coins themselves. Perhaps the cause was a combination of all of these events.

The fact is that this phenomenon is seen almost exclusively on the P mint coins from 1941 through 1950, and is especially noticeable on the coins of 1942, 1944 and 1945, 1948 and 1949.

Orange-Peel

Many Jefferson Nickels of the 1980's exhibit a marked "orange-peel" appearance. Surfaces of these coins show a mottled wear, no doubt from improper die heat treating and extended die use that resulted in severe die erosion. This orange-peel wear is common to 1980 through 1986 coins.

"Halo Effect"

A curious luster contrast is frequently encountered on San Francisco minted Jefferson Nickels through 1954. This luster contrast is circular in nature, and is apparently the result of uneven die wear or die polishing or even a combination of the two. Often heavy metal flow lines are present, especially around the letters and devices.

The "halo effect" is especially common on 1939 through 1941 issues. Curiously it seems to be absent on early issues with the reverse design of 1938 (the uneven weak steps) and on wartime silver alloy nickels. Thus this unusual die wear appears exclusive to copper-nickel coins with the reverse of 1940 design.

Prooflike Surfaces

Prooflike mirror surfaces on coins is ascribed to new dies or polished used dies. Prooflike coins are encountered most often on San Francisco Mint products in all coin series. (This seems to be a tradition of the San Francisco Mint, extending back to its earliest days) Many of the S mint prooflike coins show evidence of wear, with some details weak as a result of polishing. Under magnification, polishing scratches in the die, appearing on the coin as raised lines, are often present.

Prooflike Jefferson Nickels are less common to downright rare on nickels issued in Philadelphia or Denver in the 1930's and 40's. Such coins are more common after 1950 through 1970. From 1971 through 1986, Jefferson Nickels can be very flashy, however actual prooflike surfaces are not common, although D-mint prooflike reverses begin to appear in the 1980's, apparently the result of reused reverse dies from proof coinage.

Philadelphia and Denver mint prooflikes are usually uniform in surface appearance and seldom show die polishing lines.

The depth of prooflike surface varies greatly. Most are actually only semi-prooflike, however very deep mirror fields are sometimes encountered, even with cameo contrast.

Carbon and Corrosion Spots

Copper-nickel coinage is susceptible to the formation of small dark spots commonly referred to as "carbon spots." Actually dark spots can have two causes, but the result from both causes is usually referred to as a carbon spot.

True carbon spots are just what the name implies, a speck of carbonized material imbedded in the surface of the coin. This is thought to be caused by a contaminant that was imbedded in the surface of the planchet and subsequently reduced to a cinder by the heat of the annealing oven.

True carbon spots can be identified under high magnification by their characteristic of being at or below the surface of the coin. True carbon spots usually stay the same over time, and they are pretty much non-removable.

Corrosion spots on copper-nickel coins often have the appearance of true carbon spots when viewed under low magnification. High magnification (30X or greater) will reveal a raised mound of molten-like material surrounded by a dark area. These corrosion spots start as a speck of paper or cardboard dust. The combination of the speck, moisture in the air and the coin's metal surface result in what's called an oxygen cell. In simple terms a tiny electrical battery is created on the coin, and the resulting electrolytic action causes a dark corroded depression to form on the coin. The spots start out green, progressing to brown to black. They usually grow in size, and can be as large as 1/4 inch in diameter.

Corrosion spots can be prevented by keeping the coins away from acidic paper or cardboard. Once the spot forms, there is nothing you can do, short of carefully removing the offending material using a steady hand holding a toothpick or pin under high magnification in good light.

DESIGN CHANGES

Reverse Change: 1939

The reverse design introduced in 1938 featured weak, wavy steps on Monticello. For some unknown reason, the Mint decided to prepare a new reverse featuring a thickened porch and sharply defined steps. The new reverse was used on all regular issues through 1970, although it was slightly changed in 1967.

Comparison of the original 1938 reverse on the left, with the new reverse design introduced in 1939.
The old reverse, shown at left is now referred to as "Reverse of 1938."
The new reverse shown at right is now referred to as "Reverse of 1940."

The new reverse change was apparently implemented in early 1939. Philadelphia minted coins with the old reverse are scarce, indicating few dies were used having the old design. These old design dies may have even been carryovers from the previous year. The old reverse is more common on Denver coins, and still more common on San Francisco Coins.

It is possible that the Philadelphia Mint shipped a large number of old reverse dies to the branch mints, insuring uninterrupted production while the new reverse was on trial in Philadelphia where most of the mint's technological resources were centered. Philadelphia would have had the expertise present to address any problems that arose.

Close-up of the steps area of the original reverse of 1938 design

Close-up of the steps area of the new reverse, called "Reverse of 1940."

The design was a success, and new dies were subsequently sent to Denver and San Francisco.

That there are two different reverse designs for 1939 is a little known fact to numismatists. The Jefferson Nickel thus shared the same fate as every nickel five cent piece before it – the original design saw a change shortly after its introduction.

Terminology for the 1939 change

To avoid confusion with the wartime alloy change, which was long improperly referred to as "Type I" and "Type II" I proposed in the first edition of *The Jefferson Nickel Analyst* that the reverse changes be designated as "Reverse of 1938" and "Reverse of 1940." Variety specialists proposed the term "Variety 1 steps" and "Variety 2 steps" or just Variety 1 and Variety 2. Others proposed using "Type 1 steps" or "Type 2 steps." The PAK Full Step Nickel Club preferred to use the Variety 1 steps/Variety 2 steps designations. At the time, I was unable to get any definitive definition of what the difference was between the terms "type" and "variety." Reference books and articles differed, and so did the various specialists and experts. So I decided to avoid the controversy by using the "reverse of" designation.

The question of what to call the step difference was apparently settled by Walter Breen who listened to the opinions and decided to use the Reverse of 1938/Reverse of 1940 designations when he published his *Encyclopedia of U.S. And Colonial Coins.*

Reverse Change: 1942

When the composition of the Jefferson Nickel was changed to a silver alloy, the Mint signaled the change by suspending use of the small mintmark at the right of Monticello, and instead added a large mintmark above the dome on the reverse. For the first time, the Philadelphia Mint used the mintmark "P." I doubt this change was done for the public's benefit, but was a means for the Treasury Department to differentiate the new versus old alloy for whatever purposes they might have had, such as a means to easily identify the coins should they need to be withdrawn from circulation at a later date. (Obviously they didn't foresee that the coins would turn black in circulation and thus stick out like a sore thumb!)

The change occurred in 1942, so both alloys were issued by Philadelphia while Denver issued only copper-nickel coins and San Francisco only silver. Although the silver alloy was authorized through 1946, the composition and reverse design reverted to the previous type for 1946 and later years.

A 1942 San Francisco striking in a silver alloy with the old mintmark style is known. (see varieties section) Copper-nickel strikings are known for the 1943P and 1944P. A 1943P is known struck on a steel planchet. A sliver alloy 1946 has been reported. Counterfeit 1944 nickels are known, struck on copper-nickel planchets without the "P" mintmark.

Obverse Modification: 1957 and 1958

The star between "LIBERTY" and the date was enlarged and sharpened on all 1957 and 1958 issues. The design reverted to the previously used small star in 1959. No reason for this change has ever come to light. There was a report by a collector that he had found some 1957 coins with the small star, however this has never been verified.

Obverse Modification: 1966

Felix Schlag's initials FS were added to the obverse at the base of Jefferson's bust in 1966. This change has been permanent since then, despite any subsequent changes to the obverse.

Two proof coins were purportedly struck in 1966 and presented to Mr. Schlag at the Chicago American Numismatic Association convention. Research by Theodore Riviere revealed that in fact the coins were apparently not proofs, but were simply the first two coins struck from regular issue dies. Mr. Riviere was backed up by the late Frank Annunzio, Chairman of the Congressional committee overseeing coinage issues, who wrote to Riviere: "Mr. Schlag did not receive any proof nickels at the American Numismatic Association Convention in 1966. He was presented with the first two nickels struck that bear his initials. These were regular business strikes and were not proof coins."

Reverse Change: 1967

The reverse of 1940 was slightly modified for the special mint set issues of 1967, and used for all business and proof strikes from 1968 through 1970. The door and window lines within the porch and the steps were strengthened. The steps appear to curve slightly from top to bottom into the field above the word "MONTICELLO." This modification significantly increased the possibility of finding six step coins, at least on the proofs.

Reverse Modification: 1967 The window and door details are stronger and the steps curve slightly from the porch toward the field above MONTICELLO

Reverse Change: 1971

1971 saw the first major change in the Jefferson Nickel reverse since its inception. By 1970, the reverse of 1940, even with its modification in 1967, was showing its age as far as the circulation strike coins were concerned. The Mint decided to do a major overhaul of the reverse.

The details of the building were considerably sharpened. The lines and details of the windows, balconies, pillars and lintel above the porch were all strengthened, resulting in the best looking rendition of Monticello to date. The steps were changed so that the top porch step is about as thin as the other steps. The step area itself is parallel to the field, lacking the curve associated with the 1967 modification of the reverse of 1940.

All of the 1971 issues had the new Reverse of 1971. The coins with the new reverse appear much flashier than the previous design, and it appears the mint was using different die and/or planchet preparation to get this effect. The sixth step, although thinner than the steps above, is frequently found on coins minted during the first several years.

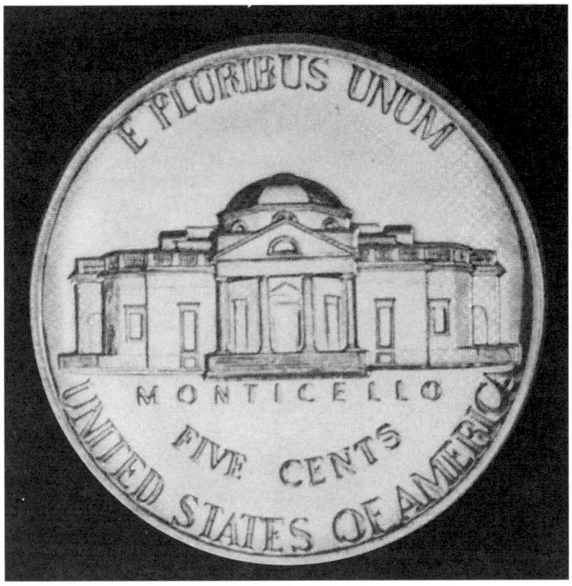

Reverse of 1971
Overall details on building were strengthened

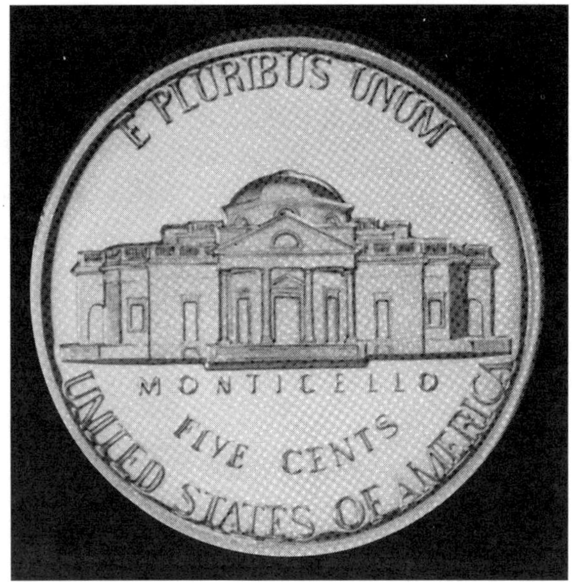

Reverse of 1977
Overall details on building slightly modified.

See also the close-up photographs on the next page.

Obverse Modification: 1977

Details of Jefferson's hair running left to right across the top, right were strengthened slightly. The cheekbones appear to be slightly softened.

Reverse Change: 1977

The details of Monticello were again modified. The left balcony details are softer. The arch-topped windows in the dome and lintel are softer. There are more line details in the windows and door on the porch. The step detail is a bit deeper.

Unlike the change in 1971, the 1977 reverse change did not increase the availability of six step coins. Although 6 step coins can be found, their frequency of appearance is not as high as one would expect for a new reverse master hub. In fact, even in the first year, six step coins are rare on the Philadelphia Mint issues. The reverse of 1977 was produced through 1986.

Close-up view of the Reverse of 1971 above, Reverse of 1977 below

Obverse Modification: 1982

Gradual spreading out of the obverse lettering led to a new obverse in 1982. The change was implemented during 1982, resulting in two different obverses for the year. A new master hub was prepared, with the lettering being both sharpened and moved further away from the rim. The gradual wear and spreading of the old master hub had resulted in loss of detail in the lettering. This is especially noticeable on the early 1982 issues.

Walter Breen refers to the new obverse as "Obverse of 1982" and to the old obverse as "Obverse of 1981." Actually, the old obverse, following this nomenclature, should properly be called "Obverse of 1977" since that is the first issue year with that reverse.

Reverse Change: 1982

The reverse of 1977 was retooled in 1982, apparently for the same reason as the obverse – wear and spreading on the master hub.

The change was first noticed during 1982 by nickel collector Dwight Stuckey, who reported the change to the numismatic press. The change apparently took place about May of 1982. Later, collector Nick Carl observed that the new reverse was actually modified in two stages.

Changes to the reverse included moving the lettering away from the edge, sharpening the steps and sharpening the architectural details of the building. The difference is especially noticeable on the dome. The two curved roof lines that run to the tip of the dome are very weak on the old reverse but very sharp and raised on the new reverse. The trim at the edge of the dome roof exhibits sharp, distinct lines, while the old reverse has these details missing or at least well worn. This change was placed in production, but for some unknown reason, the Mint decided to modify the reverse further.

The second modification to the reverse took place in the triangle above the porch, where the trim was sharpened with the addition of a prominent groove in the center of the trim, running all the way around the triangle and the half-round window. Most collectors make no distinction in these two stages for the purpose of identifying the new reverse, but instead rely on the changes in the dome to easily spot the change. The differences are best observed by inspecting 1982 nickels. The contrast between those with the old, worn reverse is easily seen when viewed side-by-side with the new, sharp reverse.

One caution however – both types of reverses come struck from worn working dies, exhibiting erosion and orange-peel surfaces. One has to look at the building detail, not the overall strike appearance, to judge if the coin was struck with the old or the new reverse.

Die Combinations for 1982

It seems that the Mint used both old and new obverse and reverse working dies in all four possible combinations. Since most collectors are not collecting the 1982 issues by obverse and reverse combinations, little study of the rarity of the combinations has been done. The new reverse is more common than the old, and full step coins are much more common on the new reverse, with six step coins appearing with some frequency.

Changes after 1982

The Mint has periodically strengthened the details of both the obverse and reverse since 1982.

Both obverse hair and reverse building details were reengraved for the 1987 coinage. The result was that six step coins are now commonly seen on all issues after that date.

The obverse hair was again sharpened in 1991 and again in 1993.

Since 1993 there have been periodic enhancements to the details, so that Jefferson Nickels of the 1990's and 2000's have sharp details. Six step coins are the norm.

GRADING

Welcome to the most controversial area of numismatics. In the case of Jefferson Nickels, the first question for most of us is: Is the nickel uncirculated? The next one is: How nice of an uncirculated coin is it with regard to eye appeal and overall strike quality? Finally, is the coin a full step coin?

I should point out that grading is subtle, and is as much art as science. The best teacher is experience with original coins so you learn what normal uncirculated Jefferson Nickels look like. This "look" varies from year to year and from mint to mint. The only way to become grading proficient is to look at many different coins. It is also beneficial to check the descriptions in the various grading guides. If you only check one of them, I would suggest *Official ANA Grading Standards for United States Coins* published by The American Numismatic Association.

Is the coin uncirculated?

Use incandescent lighting. Look for rubbing on the high points. On the obverse check the eyebrow, hair and cheekbone. On the reverse, check the dome, lintel (porch triangle), pillars and steps. Using your eyes without magnification tilt and rotate the coin and look for a contrast of luster between the high points and the rest of the coin. If there is a contrast, carefully inspect for rubbing. At this time, use up to 7X magnification if necessary.

Sometimes abrasion will be seen on the high points, which renders the coin NOT uncirculated, even if the coin came from an original roll. The presence of bagmarks (nicks and small scratches) are OK. Beware of unnatural color or luster which might signal a counterfeit, whizzed (wire brushed) or cleaned coin.

How nice is it?

Here is where the real controversy begins. Although general guidelines have been published defining the various grades of uncirculated, it is probably impossible to consistently interpret grade differences, and when this fact is combined with personal preferences and experience, a nickels grade can be seen differently by different collectors or dealers.

Even the commercial coin grading services vary in their standards, and so we see the phenomenon of the same coin, in, say MS65, sold for different prices depending on the grading service that assigned the grade.

I could insert my own opinion, but instead I will leave it to the numismatic fraternity to continue to argue about it. One thing is certain – ultimately the determination of grade and price is between the buyer and seller, and if both are experienced, ultimately the price is the indicator of how nice the coin is.

Here are some general guidelines for circulation strike Jefferson Nickels.

Uncirculated or Brilliant Uncirculated (BU) or Mint State 60 (MS60)

The coin will be quite ugly with many bag marks, or an extremely rough surfaces or poorly struck, or having excessive die erosion. If the coin is very dark or has ugly toning, or has excessive black spots then the nickel is MS60.

Choice Uncirculated or Choice BU or MS63 or MS64

A fair number of bagmarks present or a rough surface or weak strike. Darkness or non-spectacular toning or some spotting may be present. The difference between MS63 and MS64 can be subtle and experience is the best teacher.

Gem Uncirculated or Gem BU or MS65 or MS66

A very nice nickel with minimal bagmarks, minimal surface roughness and decent strike. Luster or toning should be eye appealing. A MS66 will be a very nice coin, and for many dates and mints, finding a MS66 is difficult to near impossible.

Superb Uncirculated, Superb BU or MS67 or MS68 or MS69

A really great looking coin with hardly any tiny marks, no roughness, fully struck. May have truly spectacular toning (this is rare for a Jefferson Nickel) or phenomenal luster. Very few Jefferson Nickels grade this high, even for early dates. Circulation strike MS68 or MS69 coins are practically non-extant.

Importance of "Strike"

General grading as outlined above only casually utilizes the depth of the strike of Jefferson Nickels. For the Jefferson Nickel specialist, the strike quality is important, not only when considering the steps on the reverse, but the other obverse and reverse features as well.

Strike quality is determined by striking pressure, coining press speed, die wear, presence of foreign material such as lint and grease on the dies and hardness of the planchet. Jefferson specialists naturally search for the perfect coin so a fully struck coin is important.

The term "weak strike" implies that a coin with missing detail was somehow produced with less than full die pressure, but this is not the normal situation. Most likely, die state, die wear or filled dies are the culprit.

Even brand new working dies may be already worn, the result of being prepared from worn working hubs and/or worn master dies and /or a worn master hub. In fact, progressive wear on the master hub for the Reverse of 1940 resulted in more and more loss of detail until the hub was ultimately retired in 1970. For this reason, nickels of the 1950's and 1960's are never found with as robust an appearance as earlier dates.

Quality of obverse strike can be judged by looking at the highest point, the hair curl by Jefferson's ear. It seems that this area of the die was prone to filling with greasy dirt, therefore the curl may not be fully struck up. The phenomenon became more common beginning in the 1950's, but seems to have pretty well disappeared after 1970.

On the reverse, the details of Monticello suffer, particularly in the details of the windows, porch and steps. Strike quality can be very uneven: I have seen coins where all the building details are poor but the steps are sharp!

On both sides, worn lettering may be present, the result of general die wear. Severely worn lettering is often seen on issues of the 1960's and 1980's. Some of this lettering even appears to be doubled, but this is often a phenomenon of the die wear, not a true doubled die.

Field erosion, usually appears as the "halo-effect" discussed earlier in this book, or wavy-ness or an orange-peel effect.

In searching out fully struck Jefferson Nickels, experience will be your guide as to what is the best you can reasonably expect.

A fully struck 1954S – the ideal to strive for!

Determining "Full Steps"

The term "full steps" means that at least five steps are complete on Monticello. The step count always includes the porch deck counted as a step. As designed, Jefferson Nickels have six steps. The number of steps, or the completeness of the steps that show, varies on the individual coin. Many issues are difficult to impossible to find with even five complete steps, and most issues before 1971 do not have six steps on the working die. It is possible to complete a set of regular issue Jefferson Nickels with five steps, but it is impossible to do this for six steps since many issues do not exist with all six steps. For this reason, the term full steps has come to mean that at least five complete steps are present. Complete six step coins command a premium, except those after 1990, and are referred to as "six full steps."

General Guidelines

Use a 7 to 10X, and not higher than a 20X magnifying glass to inspect the steps on the reverse. A full step coin must have complete steps and those steps must be free of deep nicks or abrasions. Nickel collectors seek steps that are clean and sharp, and anything less is not worth as much or as desirable. Any bagmarks in the step area, usually showing as nicks and abrasions, must not be deep enough to cut all the way into the depressions between the steps.

Jefferson Nickels before 1970 are prone to show what at first appears to be complete steps, but upon careful inspection, bridges can be seen between the raised lines of some of the steps. The bridges are usually no longer than 1/8 inch. The presence of even one of these bridges means that the coin is not full step.

Usually, the weakest area of the steps on nickels dated before 1971 is under the third pillar counted from the left. This area is opposite the high point on the obverse, thus the planchet metal often did not readily flow into that area. It is not uncommon to find a nickel with the first three steps complete, and the fourth and fifth steps missing in a pinhead area just below the third pillar. Design changes after 1970 seem to have eliminated this problem.

Specific Guidelines: Full Step Reverse of 1938

The original design, produced in 1938 and part of 1939, had weakly defined steps. The steps appear mushy and not clearly defined, even on proofs. The steps are uneven and wavy and the indentations between the steps vary in depth. There are many bridges present between the raised portions of the steps.

The best way to grade the Reverse of 1938 is by overall definition of the step area. The steps should be overall reasonably struck with most of the definition present, especially at the left.. A five step coin will have a defined, although thin, raised line all the way across at the fifth step. The line must be defined below the third pillar. If you begin looking from the left, you will probably see the fifth step clearly. As you move from left to right, the step will get thinner, and will tend to merge with the fourth step. For five full-steps the fifth step must not merge with the field below the third pillar, but instead must have some definition.

A six step coin will have a very thin sixth step going all the way across. Even on proofs, six step coins are seldom found.

Properly defining a five step Reverse of 1938 nickel is difficult due to the nature of the design. A good way to familiarize yourself with this reverse is to inspect a 1938 proof coin. All of them have at least five steps, and you will be able to see what fully struck five step coins look like. Circulation strikes will have the same characteristics, but not as sharp a detail.

A five full step 1938 nickel

Specific Guidelines: Full Step Reverse of 1940

The reverse of 1940, used on Jefferson Nickels through 1967, and with modification 1968 – 1970, has straight, sharply defined steps. This definition, however progressively deteriorated as the master hub wore. Thus the steps found on earlier dates will be sharper and deeper than on later dates.

A five step coin must have four defined indentations between the raised portions of the steps and the raised lines and the indentations must be complete from left to right. If the raised portions bridge together at any point, then the coin does not have five full steps. In some cases, all five or six steps will be present, but there will be a small bridge somewhere within the steps. Such a coin is not a full step coin.

Bagmarks and struck in marks, depressions, etc. are commonly seen on this reverse, especially on later issues. These detractions must be minor and not cut deeply into the steps. Even when these detractions are minimal, they do decrease the desirability of the coin as far as a Jefferson specialist is concerned. Thus the most desirable nickel is one with sharp, clean steps. Such coins are easier to find in earlier years, and this fact should be taken into account when considering nickels after the mid 1940's. Many later dates are difficult to find with perfect steps.

Six step coins should have a sharply defined last step going all the way across. On later issues the sixth step is thinner than the preceding steps.

A full step example of the reverse of 1940.
Almost half of the sixth step is showing.

Specific Guidelines: Full Step Reverse of 1971

The reverse of 1971, used through 1976, possesses sharp though thin steps. This reverse does not seem prone to bridges between the steps, and the only problem, besides the steps not being there at all, is presence of bagmarks. Because of the thin steps, cuts and nicks easily spoil an otherwise full step coin. The sixth step when present is always very thin.

Specific Guidelines: Full Step Reverse of 1977

The reverse of 1977 was used into 1982. The steps are somewhat thicker than the reverse of 1971. As with the reverse of 1971, the major problem is presence of five steps or more, and marks in the steps. The steps are not as sharp after 1979, due to master hub wear. Also, die wear became a major problem after 1979, thus many coins will show worn lettering and orange-peel effect. The sixth step when present usually has the same thickness as the preceding steps.

Specific Guidelines: Full Step Reverse of 1982

The reverse of 1982 was used until 1987, after which it was reengraved, with subsequent touchups and re-engravings since. The steps are sharper than the reverse of 1977, and the modifications of 1989 and later made them even bolder. The 1982 through 1986 issues show progressive master hub wear resulting in more poorly defined steps. After 1986, the Mint maintained the hub often enough that five and six step coins are easy to find. This is particularly true for the issues of the last decade.

Partial Step Coins

Although most Jefferson Specialists want only five to six full step nickels, some dates are so rare in that condition that one has to settle for a partial step coin.

There are two methods in use to describe partial step coins. One simply uses a verbal description while the other divides the step area into quarters and then details what steps are full in each quarter.

Description Method

The step area is simply described as far how close to five full steps the coin is. A coin with five steps save for a small soft area bridging two steps would be described as "almost 5 steps." If the coin had partial steps, the approximate fraction of the steps showing is described; for example: "1/2 steps" or "3/4 steps."

None of the third-party grading services recognize less than five step coins using simple descriptions as outlined above.

An "almost 5 step" 1938 nickel with softness under the 3rd. pillar

An "almost 5 step" 1965 nickel has an incomplete 5th step at the 3rd. pillar

Dividing Steps into Quarters

The PAK Full Step Nickel Club devised a method of describing partial step coins by dividing the step area into quarters. On a six step coin, each quarter would have 6 steps and the total for the four quarters is 24. If a nickel has incomplete steps in a given quarter that area is not counted. Thus a partial step coin will have less than 24 quarters. The total of the quarters was divided by four to get the step count.

Although the PAK method gave an overall step count it did not provide a detailed description of where steps were missing. To correct this, Darrel Crane of the Full Step Nickel Club revised the PAK method to number the steps in each quarter 1 through 6. The steps are described by using four separate numbers designating how many steps are complete in each quarter. Thus, a "6666" would describe a full six step coin. A "5555" would be a five full step coin.. Typically partial step quarters are present, so descriptions such as "6655" for a 5 1/2 step coin, or "6545" for an almost 5 step coin would be common.

SEGS Grading Service recognizes and uses this system when they grade Jefferson Nickels.

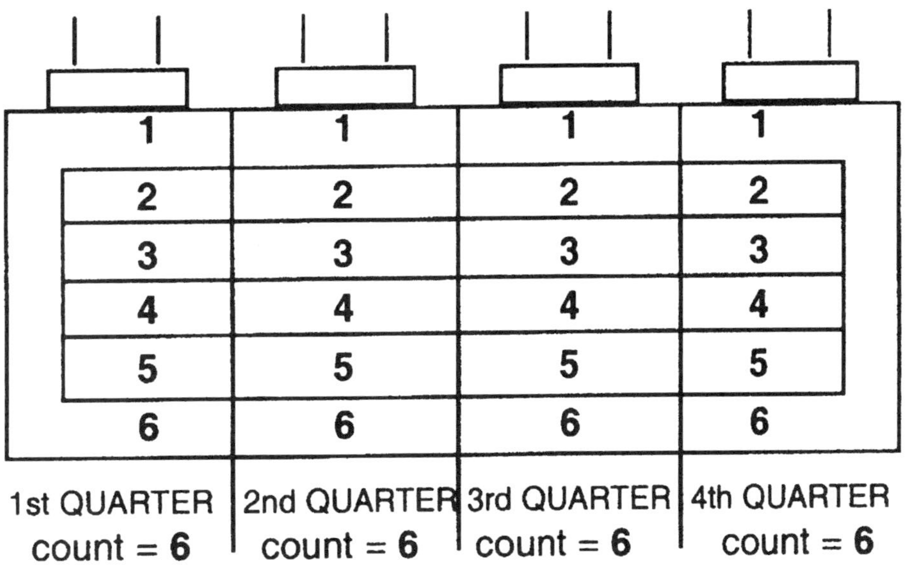

FSNC Quarter Step Grading System

RARITY: Uncirculated Rolls

The availability of BU rolls of Jefferson Nickels varies, dependent upon popularity, price, and release of hoards, long-held material or estates. In the 1950's and 1960's, collecting modern coins by the roll was popular, and many coin dealers specialized in that area, buying and selling rolls of cents through half dollars dated after 1933. This area of collecting had evolved to a speculative fever by the early 1960's, only to be cut short by the coin shortage of 1964, the elimination of silver from coinage, and the assault of the US Mint on coin collectors, blaming them for the coin shortage.

The popularity of roll collecting was evident by the mid-1940's, accelerating in popularity through the 1950's and culminating in the mid-1960's. Shortly after, roll collecting suffered a precipitous drop. By 1970, there were few roll collectors, and BU roll dealers switched to other numismatic areas or retired. By about 1975 there were only three coin dealers who regularly advertised BU rolls, and two of them specialized only in rolls after 1964.

Thus the general statement can be made that early Jefferson Nickel rolls were not collected to the extent that later rolls up to 1964 were, and that rolls after 1964 were not saved to the extent that earlier rolls were.

In discussing their availability, BU rolls of Jefferson Nickels can be divided into several groups: pre-wartime, wartime silver, post war to 1964 and 1965 to date.

Pre-wartime BU rolls 1938-1942

Generally the pre-war copper-nickel issues are scarce to rare. Roll collecting was not that popular at the time and the US was in economic turmoil at the end of the great depression. Some typical "new issue hoarding" occurred with the release of the Jefferson Nickel in 1938, but this was more concentrated in the lower mintage branch mint issues. In terms of actual availability, it is easier to find rolls of 1938D nickels than 1938S and it is much easier to find either D or S rolls than it is to find P mint rolls. The 1939 P mint rolls seem to be as scarce as the 1938. Of course the 1939D and 1939S rolls are seldom seen, in keeping with their low mintage. It is rumored that a bag of 100 rolls of 1939D coins was split up in the 1960's.

The 1940 and 1941 P mint coins are more common than 1938 or 1939, but even these two dates, the most common of the pre-war nickels, are scarce when viewed against the greater availability of post 1942 rolls. Both D and S mint rolls of 1940 and 1941 are quite scarce, with the S mint rolls being scarcer, especially the 1941S.

Years ago I was laughed at when I said that the 1942 copper-nickel issue was rarer that the 1942P silver nickel, especially in gem condition. Current price quotes show both issues being valued about the same, after many years of the 1942 copper–nickel rolls valued at 1/10th the value of the silver roll. In fact, 1942 "type 1" rolls are very hard to find, being scarcer than 1940 or 1941 P mint rolls. The 1942D is in a class by itself. Despite its relatively high mintage, the 1942D is a rare coin by the roll. My own experience is that it is in fact rarer than either the 1939D or the 1939S.

The condition of early BU rolls is generally very good, except for the S mints with the reverse of 1940 – they often have a washed out, almost circulated appearance. The situation got progressively worse through 1941, thus gem rolls are quite rare, and most S mint rolls will not be as frosty or flashy as the P and D mints. The P's and D's are usually very nice, with the D's being better struck and more flashy than the P's. Some P and D mint rolls show spotting – check before you buy!

I have never heard of any hoards of the pre-war copper nickel issues.

Wartime Silver Rolls

The silver alloy nickels issued from 1942-1945 had an entirely different, more appealing look than the earlier nickels. This appearance no doubt encouraged saving of these coins, since as a general statement the wartime silver nickels are much more available than the earlier coins, even for the 1942 issues.

Although all dates and mints have been readily available by the roll, some issues are scarcer than others. My experience has been that the 1944D roll is the hardest to obtain, followed by the 1944P and 1945P. Next is scarcity seems to be the 1942P and S and 1944S. The 1943D roll is not as scarce as its mintage would indicate. The 1943P, 1943S and 1945S are the most common.

Hoards of wartime silver nickels do exist. In the 1970's I was told that I could buy a bag of 100 rolls – or even multiple bags – of any date or mint, provided the price was right. In the 1980's, when wartime BU rolls were high priced, multiple BU rolls could be typically purchased at any large coin show.

A large hoard was disbursed in the early 1990's, consisting of about 500 rolls of all dates and mints. The gem coins were cherry picked from the rolls and sold as individual coins after being graded by Numismatic Guaranty Corporation. A hoard of 518 BU rolls of 1943P nickels was discovered in 1995.

BRILLIANT UNCIRCULATED ROLLS

Dates	Cents	Nickels	Dimes	Quarters	Halves
1935	44.00	125.00	95.00	265.00	125.00
1935D	115.00	275.00	1050.00	1650.00	510.00
1935S	185.00	250.00	385.00	1550.00	750.00
1936	28.00	125.00	84.00	270.00	125.00
1936D	80.00	135.00	560.00	—	235.00
1936S	115.00	135.00	340.00	1250.00	385.00
1937	28.00	150.00	95.00	220.00	150.00
1937D	50.00	140.00	210.00	375.00	675.00
1937S	75.00	160.00	205.00	2300.00	580.00
1938	52.00	115.00	85.00	1700.00	350.00
1938D	125.00	285.00	330.00	—	—
1938D Buf	—	125.00	—	—	—
1938S	165.00	360.00	200.00	950.00	—
1939	25.00	47.00	90.00	240.00	180.00
1939D	215.00	2250.00	80.00	330.00	180.00
1939S	63.00	580.00	330.00	1075.00	290.00
1940	23.00	32.50	90.00	275.00	95.00
1940D	45.00	55.00	85.00	1600.00	—
1940S	33.00	84.00	85.00	225.00	175.00
1941	30.00	30.00	58.00	82.00	85.00
1941D	52.00	52.00	125.00	180.00	215.00
1941S	53.00	58.00	95.00	220.00	465.00
1942	17.00	63.00	50.00	85.00	70.00
1942P	—	320.00	—	—	—
1942D	17.00	475.00	63.00	100.00	210.00
1942S	185.00	140.00	130.00	710.00	215.00
1943	13.00	65.00	53.00	45.00	75.00
1943D	25.00	165.00	55.00	125.00	162.00
1943S	50.00	55.00	65.00	230.00	225.00
1944	9.00	100.00	45.00	40.00	72.00
1944D	13.50	80.00	48.00	140.00	95.00
1944S	17.00	130.00	53.00	95.00	105.00
1945	13.00	112.00	60.00	30.00	73.00
1945D	17.50	59.00	45.00	78.00	75.00
1945S	21.00	46.00	46.00	60.00	97.00
1946	8.00	17.00	19.00	40.00	95.00
1946D	19.00	35.00	44.00	120.00	200.00
1946S	19.00	75.00	47.00	172.00	118.00
1947	29.00	19.00	52.00	65.00	120.00
1947D	15.00	41.00	56.00	72.00	110.00
1947S	39.00	82.00	86.00	165.00	—
1948	15.00	18.00	125.00	36.00	185.00
1948D	14.00	60.00	72.00	62.00	93.00

Dates	Cents	Nickels	Dimes	Quarters	Halves
1948S	75.00	100.00	77.00	58.00	—
1949	26.00	47.00	440.00	530.00	630.00
1949D	23.00	52.00	150.00	125.00	195.00
1949S	84.00	155.00	565.00	—	380.00
1950	14.00	175.00	78.00	48.00	315.00
1950D	10.00	850.00	80.00	48.00	185.00
1950S	41.00	—	500.00	150.00	—
1951	30.00	45.00	62.00	33.00	75.00
1951D	7.00	75.00	46.00	29.00	170.00
1951S	53.00	325.00	445.00	240.00	165.00
1952	28.00	21.00	22.00	29.00	77.00
1952D	6.50	195.00	22.00	23.00	33.00
1952S	35.00	48.00	115.00	134.50	140.00
1953	9.50	16.00	44.00	72.00	230.00
1953D	4.75	16.00	16.00	20.00	25.00
1953S	22.00	46.00	43.00	46.00	80.00
1954	39.00	12.00	13.00	17.00	33.00
1954D	6.50	8.00	13.00	26.00	20.00
1954S	19.00	23.00	46.00	55.00	53.00
1955	6.75	130.00	120.00	55.00	185.00
1955D	5.50	15.00	76.00	175.00	—
1955S	35.00	—	48.00	—	—
1956	5.00	10.25	10.50	15.00	75.00
1956D	2.25	4.50	9.50	18.00	—
1957	4.75	13.50	8.50	14.00	42.00
1957D	1.75	3.50	9.50	13.50	16.50
1958	5.00	35.00	42.00	66.00	70.00
1958D	1.75	3.50	7.50	13.00	16.50
1959	1.75	12.50	8.50	17.00	38.00
1959D	1.25	2.80	7.50	13.00	25.00
1960	1.25	4.25	9.75	13.50	35.00
1960-P-SD	430.00	—	—	—	—
1960D	1.25	3.25	6.75	13.50	16.50
1960-D-SD	19.00	—	—	—	—
1961	1.30	3.75	12.00	13.50	27.00
1961D	1.10	3.00	6.50	13.50	18.50
1962	1.25	3.50	8.00	13.50	25.00
1962D	.85	3.00	6.25	13.00	13.00
1963	.85	3.00	6.25	12.50	14.00
1963D	.75	2.60	6.00	12.50	12.50
1964	.70	2.60	5.75	11.75	11.75
1964D	.70	2.60	5.75	11.75	11.75

A price guide for BU nickel rolls published in 1965

Post War-Time Rolls 1946-1964

BU roll collecting and hoarding really entered its heyday after World War II. The craze particularly affected S mint coins which were widely saved due to their generally lower mintage.

All of the 1946-1964 issues are actually common as BU rolls. Judging from their appearance in dealer shops and ads and at coin shows, one might conclude that these rolls are hard to find, but the truth is that this material does exist in quantity.

The reason one does not see it that often is that the price is not high enough in many cases to bring it out. Understand that many old time collectors are "buried" in this material, purchased at very high prices during the BU roll craze of the 1960's. You can understand that a collector who purchased a roll of 1950D nickels for $1000. or a roll of 1955 nickels for $150. might be a reluctant seller at $250. and $10.00 respectively.

This material is periodically available when someone throws in the towel or when a collector dies and his estate blows out his formerly high priced BU rolls. When I was actively purchasing large quantities of BU nickel rolls in the 1970's and 1980's, I was able to find the material readily, especially rolls dated after 1950. Once, I was offered and purchased dozens of original 1949S rolls. I also purchased some rolls from a 100 roll lot of 1950 nickels. At one coin show, a dealer had several hundred BU 1950D rolls available. I routinely purchased large lots of nickels dated 1956-1964 – but had to dump them into circulation because they were almost un-saleable. I could have purchased even more if I was willing to pay a higher price, or if I was willing to travel to the place where a collector or dealer had the material stored – it wasn't worth the collector's or dealer's time otherwise. I have no doubt that much of this material still exists, waiting for the right time or price to enter the market.

Within the "commonness" of the post-wartime rolls, some rolls are less common than others. Among the P mints, the 1949 and 1950 are scarcer than the others. Of course, gem, brilliant rolls of the 1952-1955 P mint issues are downright uncommon to rare – most rolls encountered are horribly dark and weakly struck.

Among the D mints, the 1949D, 1951D and 1952D rolls are harder to buy than the others. All of the D mint rolls between 1946 and 1952 are rarer than the 1950D roll, and thus are comparatively underpriced. A 1000 roll hoard of 1946D was broken up in 1972!

Most S mint rolls are more common than P or D mint rolls, with the

1946S being most common among the pre-1951 rolls. One large hoard of 736 BU 1946S rolls was discovered in 1995! The 1947S roll is less common than the 1948S, and the 1949S and 1951S rolls are the scarcest of the post-war S mint run. When purchasing S mint rolls be aware that the rolls may contain many weak strikes, particularly on the 1952-1954 issues.

BU Rolls 1965 to date

The demise of the roll craze and of silver coins after 1964 led to a rapid decline in the practice of saving BU nickel rolls. For this reason, post-1964 rolls are theoretically less available than earlier rolls. I always found it easier to find and purchase rolls of 1964 nickels than 1965 nickels, and so on. Original bank-wrapped rolls dated after 1965 are seldom encountered.

Most of the available supply of BU rolls after 1965 comes from two sources: from the supply purchased and saved by a small number of collectors and dealers interested in that material, and from broken mint sets.

During the 1970's and 1980's there were ongoing promotions featuring sets of Kennedy halves and Eisenhower dollars. This created an insatiable demand for BU coins which quickly exhausted the small supply of BU rolls and bags. The only source left was original US Mint Sets, most of which could be purchased at little premium over face value. These sets were cut up for the halves and dollars. Many of the cents, nickels, dimes and quarters were dumped into circulation, but some of the coins were rolled and offered for sale. During that time I purchased many thousands of these mint set nickels, and after searching through them, released them into circulation!

Today, rolls of post 1965 nickels from whatever original source are not easy to find, or they are prohibitively expensive, a result of their scarcity. Since 1965 there has been a repeating cycle – after the year of issue, rolls are available for a time, mainly from dealers specializing in this area. After some years, these dealer hoards are disbursed, the rolls become uncommon, and the price can escalate. Witness the 1982 and 1983 years – for a time these rolls were cheap, but once the dealer hoards were exhausted the price skyrocketed, the situation made worse by the fact that there were no mint sets issued those years that could be broken up.

RARITY: Full Step Coins

When the *Jefferson Nickel Analyst* was originally published in 1979, the rarity of five full step nickels was addressed by listing the individual issues by their "frequency of appearance." This data was the result of my own experience in searching through many thousands of uncirculated coins. The data was updated in the late 1980's and published in a newsletter of The American Nickel Collectors Association.

The frequency of appearance represents the number of MS65, well struck full step nickels you could expect, on average, to find if you searched a very large number of a given issue, expressed as a ratio. A ratio of 1:50 meant that on average you could expect to find one MS65 five step nickel for every 50 coins of that date. Obviously, actual experience would depend on the particular batch of coins being searched. Also, the ratio assumed that the batch of coins being checked were original and previously unsearched. For the purposes of ascertaining relative rarity of the various issues of full step nickels, the frequency of appearance provided a useful tool.

Six step coins, due to their relative rarity for most dates before 1987, were assigned a listing using designations common, scarce, very scarce, rare, very rare, extremely rare and unverified and unknown.

The 5 step frequency of appearance and six step rarity table is presented with updated information through the year 2000. The earlier data published in 1979 is also revised where necessary based on my experience after 1979. These figures are based upon actual experience, not opinion or myth. The information given is useful in pre-determining what you may expect when searching through original uncirculated Jefferson Nickels, but keep in mind that actual results will depend on the individual batch of coins examined. Also keep in mind that unsearched, original rolls of nickels are becoming harder to find. Previously searched material will obviously yield a very different result.

Six step coins are listed using the same descriptions as previously used, but the descriptions in this edition correspond with the Universal Collectibles Rarity Scale, to which is added a few additional terms as listed below.

Unknown: probability that none exist, based on die or hub wear.
Unverified: possibility that six step coins may exist, but none reported.
Ultra Rare: 2 or 3 reported
Extremely Rare: 4 to 8 reported
Very Rare: 9 to 20 reported
Rare: 21 to 50
Very Scarce: 51 to 500
Scarce: 501-2000
Common: 2001 or more

Rarity Table for five and six full step nickels

Five step coins: frequency of appearance ratio as a MS65, well struck coin
Six step coins: listed as reported or observed rarity using UR scale

Date:	5 step	6 step	Date:	5 step	6 step
38	1:20	ex rare	48S	1:20	rare
38D	1:20	ex rare	49	1:500	unverified
38S	1:50	unverified	49D	1:20	rare
39 r38	1:1000	unknown	49S	1:150	ex rare
39D r38	1:100	unverified	50	1:150	v rare
39S r38	1:100	ultra rare	50D	1:10	v scarce
39 r 40	1:5	rare	51	1:100	v rare
39D r40	1:10	ex rare	51D	1:30	rare
39S r40	1:10	rare	51S	1:150	ex rare
40	1:5	v scarce	52	1:200	unverified
40D	1:3	scarce	52D	1:20	unverified
40S	1:10	scarce	52S	1:300	unverified
41	1:5	v scarce	53	1:500	unverified
41D	1:3	scarce	53D	1:200	unverified
41S	1:10	scarce	53S	1:5000	unverified
42	1:20	rare	54	1:200	unverified
42D	1:3	scarce	54D	1:200	unverified
42P	1:10	v rare	54S	1:3000	unverified
42S	1:20	v rare	55	1:200	unverified
43P	1:5	scarce	55D	1:200	unverified
43D	1:2	common	56	1:10	rare
43S	1:10	scarce	56D	1:150	unverified
44P	1:10	rare	57	1:10	rare
44D	1:2	common	57D	1:100	rare
44S	1:100	v rare	58	1:30	rare
45P	1:20	v rare	58D	1:10	v scarce
45D	1:5	scarce	59	1:10	rare
45S	1:300	ex rare	59D	1:20	ex rare
46	1:10	v scarce	60	1:100	unverified
46D	1:5	scarce	60D	1:5000	unknown
46S	1:30	v scarce	61	1:150	unknown
47	1:10	v rare	61D	1:5000	unknown
47D	1:5	scarce	62	1:20	unverified
47S	1:30	v rare	62D	1:2000	unknown
48	1:30	v rare	63	1:20	unverified
48D	1:10	scarce	63D	1:3000	unknown

Rarity Table for five and six full step nickels

Date:	5 step	6 step	Date:	5 step	6 step
64	1:30	unknown	83P	1:20	v scarce
64D	1:2000	unknown	83D	1:20	v scarce
65*	1:100	unknown	84P	1:30	rare
66*	1:100	unknown	84D	1:30	rare
67*	1:500	unknown	85P	1:40	v rare
68D	1:5000	unverified	85D	1:40	v rare
68S	1:500	rare	86P	1:40	v rare
69D	1:10000	unverified	86D	1:40	v rare
69S	1:2000	unverified	87P	1:2	scarce
70D	1:3000	unverified	87D	1:2	scarce
70S	1:100	v scarce	88P	1:3	scarce
71	1:3	scarce	88D	1:3	scarce
71D	1:2	scarce	89P	1:5	common
72	1:5	scarce	89D	1:5	common
72D	1:10	v scarce	90P	1:5	common
73	1:3	scarce	90D	1:5	common
73D	1:3	scarce	91P	1:2	common
74	1:10	v scarce	91D	1:2	common
74D	1:10	v scarce	92P	1:2	common
75	1:30	v scarce	92D	1:2	common
75D	1:30	v scarce	93P	1:2	common
76	1:1000	v rare	93D	1:2	common
76D	1:100	rare	94P	1:2	common
77	1:200	v rare	94D	1:2	common
77D	1:10	v scarce	95P	1:2	common
78	1:30	ex rare	95D	1:2	common
78D	1:10	v scarce	96P	1:2	common
79	1:150	ex rare	96D	1:2	common
79D	1:10	v scarce	97P	1:2	common
80P	1:20	v rare	97D	1:2	common
80D	1:10	v rare	98P	1:2	common
81P	1:30	unverified	98D	1:2	common
81D	1:20	unverified	99P	1:2	common
82P r77	1:50	unverified	99D	1:2	common
82D r77	1:30	unverified	00P	1:2	common
82P r82	1:5	scarce	00D	1:2	common
82D r82	1:5	scarce			

* regular issue (not Special Mint Set coins)

Rarity Table for five and six full step nickels
Major Varieties

Variety:	5 step	6 step
1939 double die rev (I)	1:10	unverified
1941S large S	1:10	unverified
1942D D/horiz D	1: 5	unverified
1943P double eye	1:10	unverified
1943/42P	1: 5	ex rare
1945P double die rev (I)	1:20	unverified
1946D D/inverted D	1:20	unverified
1949 D/S	1:20	unverified
1954 S/D	unknown	unknown
1955 D/S (I)	unknown	unknown

Scarce to Rare Five Full Step Nickels
Listed in order of rarity by frequency of appearance
For MS65, Well Struck Coins

unknown	1954S/D, 1955D/S
1:10000	1969D
1:5000	1953S, 1960D, 1961D, 1968D
1:3000	1954S, 1963D, 1970D
1:2000	1962D, 1964D, 1969S
1:1000	1939r38, 1976
1: 500	1949, 1953, 1967, 1968S
1: 300	1945S, 1952S
1: 200	1952, 1953D, 1954, 1955, 1955D, 1977
1: 150	1949S, 1950, 1951S, 1956D, 1961, 1979
1: 100	1939D (R38), 1939S (R38), 1944S, 1951, 1957D 1960, 1965, 1966, 1970S, 1976D
1: 50	1938S, 1982P (R77)

Date by Date Analysis: Mint State Coins

Regular Issues

Reverse of 1938

1938

Availability of Uncirculated Coins: Singles usually available, but rolls are scarce
Appearance of Uncirculated Coins: brilliant satin surfaces
Existence of Prooflike coins: periodically seen in semi to almost full prooflike
Five Step Frequency of Appearance: 1:20 Six Step Coin Rarity: extra rare

By the roll, this issue is scarcer than either 1938D or 1938S. Most 1938 singles are very nice in appearance.

1938D

Availability of Uncirculated Coins: Singles usually available, rolls periodically encountered and are not as scarce as mintage would suggest.
Appearance of Uncirculated Coins: brilliant satin surfaces
Existence of Prooflike coins: semi-prooflike coins are extremely rare
Five Step Frequency of Appearance: 1:20 Six Step Coin Rarity: extra rare

The 38D is almost always a very attractive, fully struck coin. Many have an attractive light golden toning.

1938S

Availability of Uncirculated Coins: Singles usually available, rolls periodically encountered and are not as scarce as mintage would suggest.
Appearance of Uncirculated Coins: brilliant satin or semi-prooflike, often struck from worn dies.
Existence of Prooflike coins: often found with semi-prooflike to prooflike fields
Five Step Frequency of Appearance: 1:50 Six Step Coin Rarity: unverified

Although often weakly struck, the overall appearance of 38S coins in good. The step area is usually very weak, and even five full step coins are much weaker than either 38 or 38D. Some six step coins may exist, although I have not seen or heard of any.

Reverse of 1938

1939

Availability of Uncirculated Coins: Very difficult to locate with the reverse of 1938
Appearance of Uncirculated Coins: brilliant satin surfaces
Existence of Prooflike coins: none seen – may not exist
Five Step Frequency of Appearance: 1:1000 Six Step Coin Rarity: unknown

This is a very rare coin if one considers the reverse of 1938 as a separate design type. Probably 10% of the total 1939 mintage used the reverse of 1938. I have encountered only two solid BU original rolls of this reverse, and only two other rolls that were mixed reverses. Although the frequency of appearance of five full step coins is 1:1000, the actual rarity is greater since the supply of uncirculated coins is so small.

1939D

Availability of Uncirculated Coins: Difficult to locate with this reverse
Appearance of Uncirculated Coins: brilliant satin surfaces
Existence of Prooflike coins: none seen – may not exist
Five Step Frequency of Appearance: 1:100 Six Step Coin Rarity: unverified

1939D's with this reverse probably were 25% or less of the total mintage. I have encountered only a few original rolls. The coins are almost always in choice to gem condition, but sometimes have some spotting.

1939S

Availability of Uncirculated Coins: Often seen with this reverse
Appearance of Uncirculated Coins: brilliant satin to semi-prooflike surfaces
Existence of Prooflike coins: sometimes seen, but not as often as for 1938S
Five Step Frequency of Appearance: 1:100 Six Step Coin Rarity: ultra rare

This reverse comprises about half of the 1939S mintage. In general the 1939S is as rare, and possibly rarer, than the 1939D. When one eastern US coin dealer attempted to corner the market in 1939S rolls in 1979, he was able to buy only about a dozen rolls while pushing the price from about $350. to almost $3000!

Most 1939S coins with the reverse of 1938 are choice to gem, with decent to full strikes

Reverse of 1940

1939

Availability of Uncirculated Coins: singles readily available, but rolls are scarce
Appearance of Uncirculated Coins: brilliant satin surfaces
Existence of Prooflike coins: very rare with semi-prooflike surfaces
Five Step Frequency of Appearance: 1:5   Six Step Coin Rarity: rare

Most 1939 coins made by the Philadelphia Mint have the reverse of 1940. The coins are very nice, but sometimes are spotted. BU rolls are quite scarce - once I attempted to purchase several rolls over the dealer teletype system, but only obtained one roll. To purchase that one roll, I had to steadily increase the bid price from $30. to $80.

There are several varieties of a double die reverse, one with prominent doubling. See variety section for details and photo.

1939D

Availability of Uncirculated Coins: singles usually available, rolls are rare
Appearance of Uncirculated Coins: brilliant satin surfaces
Existence of Prooflike coins: very rare with semi-prooflike surfaces
Five Step Frequency of Appearance: 1:10   Six Step Coin Rarity: extra rare

1939D's with the reverse of 1940 comprise about 3/4 of the mintage, based on my experience. Rolls do exist, and can be purchased periodically.

The 1939D with the reverse of 1940 was counterfeited in the mid-1960's. In the 1980's I was offered an entire BU roll of these counterfeits as genuine coins. To a novice, these coins are deceptive. The appearance, however is very unlike the typical brilliant, satiny 1939D. Instead, the counterfeits have a silvery gray appearance, and under magnification, the design details are not sharp. For example, the steps are not sharp as they would be on a genuine 1939D. A diagnostic of this counterfeit is a raised projection to the right of the crossbar of the A in AMERICA. The counterfeits originated in the Midwest. The same counterfeiter produced 1950D fakes that had the same appearance and diagnostic mark.

1939S

Availability of Uncirculated Coins: singles usually available, rolls are rare
Appearance of Uncirculated Coins: brilliant but smooth to prooflike surfaces
Existence of Prooflike coins: often seen with semi to full prooflike surfaces
Five Step Frequency of Appearance: 1:10   Six Step Coin Rarity: rare

About half of the 1939 mintage has this reverse. These coins are not as nice as the reverse of 38 coins are, and often are seen with an off-luster and weak strikes. Gem coins are quite rare. A dazzling 6 step coin sold for $2640. in a 1982 auction.

1940

Availability of Uncirculated Coins: singles usually available, rolls are scarce
Appearance of Uncirculated Coins: brilliant satin to frosty surfaces
Existence of Prooflike coins: very rare with semi-prooflike surfaces
Five Step Frequency of Appearance: 1:5 Six Step Coin Rarity: very scarce

Always a very nice coin, but sometimes spotted. This and the 1941 are the most commonly encountered BU rolls of the pre-wartime nickels.

1940D

Availability of Uncirculated Coins: singles usually available, rolls are very scarce
Appearance of Uncirculated Coins: brilliant satin to frosty surfaces
Existence of Prooflike coins: very rarely encountered with semi-prooflike surfaces
Five Step Frequency of Appearance: 1:3 Six Step Coin Rarity: scarce

Always a very nice sharply struck coin, but sometimes spotted. If you are lucky enough to find a roll of this issue, virtually all will have at least five steps. Six step coins are rated scarce only because rolls of 40D's are generally unavailable.

1940S

Availability of Uncirculated Coins: singles usually available, rolls are very scarce
Appearance of Uncirculated Coins: brilliant but smooth to prooflike surfaces
Existence of Prooflike coins: often seen with semi or fully prooflike surfaces
Five Step Frequency of Appearance: 1:10 Six Step Coin Rarity: scarce

1940S is often seen with an off-luster (halo-effect) and a weak strike. Gem, fully struck coins are quite rare but are very attractive when found. Some coins have a grayish, circulated appearance. Six step coins are much scarcer than the 1940D.

1941

Availability of Uncirculated Coins: singles usually available, rolls are scarce
Appearance of Uncirculated Coins: brilliant satin to frosty surfaces
Existence of Prooflike coins: very rare with semi-prooflike surfaces
Five Step Frequency of Appearance: 1:5 Six Step Coin Rarity: very scarce

Always a very nice coin, but sometimes spotted. This and the 1940 are the most commonly encountered BU rolls of the pre-wartime nickels.

Beginning with this year, granular, rough fields are sometimes encountered on Philadelphia minted coins.

Six step coins are not as common as 1940 coins.

1941D

Availability of Uncirculated Coins: singles usually available, rolls are very scarce
Appearance of Uncirculated Coins: brilliant satin to frosty surfaces
Existence of Prooflike coins: very rarely encountered with semi-prooflike surfaces
Five Step Frequency of Appearance: 1:3 Six Step Coin Rarity: scarce

Always a very nice sharply struck coin, but sometimes spotted. If you are lucky enough to find a roll of this issue, virtually all will have at least five steps. Six step coins are rated scarce only because rolls of 41D's are generally unavailable.

I have encountered some almost fully prooflike coins, however these are extremely rare.

1941S

Availability of Uncirculated Coins: singles usually available, rolls are rare
Appearance of Uncirculated Coins: brilliant but smooth to prooflike surfaces
Existence of Prooflike coins: often seen with semi or fully prooflike surfaces
Five Step Frequency of Appearance: 1:10 Six Step Coin Rarity: scarce

1941S is often seen with an off-luster (halo-effect) and a weak strike. Gem, fully struck coins are very rare but are very attractive when found. Many 1941S coins have a grayish, circulated appearance. Rolls are quite rare, and are usually substandard. I have never heard of any hoards of this issue.

Six step coins are scarcer than the 1940S.

A new, large S mintmark punch was used for later production, and on all San Francisco copper-nickel coins after 1942. It appears that less than 10 dies were used, and the 1942 large S is very rare in uncirculated. (see variety section)

1942

Availability of Uncirculated Coins: singles usually available, rolls are very scarce
Appearance of Uncirculated Coins: brilliant satin to frosty surfaces
Existence of Prooflike coins: very rarely encountered with semi-prooflike surfaces
Five Step Frequency of Appearance: 1:20 Six Step Coin Rarity: rare

The copper-nickel 1942 used to be inexpensive until someone needed a quantity of the date and drove up the price. The current price is indeed warranted, since the issue is rarer than the 1941 and is probably as rare, or even rarer than the 1942P silver alloy coin. There is one 100 roll lot reported to exist, and may have been disbursed by now.

Quality is an issue with the 1942. Although some coins exist with satin surfaces, most have rough though frosty surfaces. Finding a five full step coin is not that difficult, however finding a real gem with smooth fields is a task.

All 1942 coins were struck from a double hub. There are two doubled dies.

1942D

Availability of Uncirculated Coins: singles usually available, rolls are rare
Appearance of Uncirculated Coins: brilliant satin to frosty surfaces
Existence of Prooflike coins: very rarely encountered with semi-prooflike surfaces
Five Step Frequency of Appearance: 1:3 Six Step Coin Rarity: scarce

Long recognized as one of the key uncirculated dates, despite its relatively high mintage, the 42D is quite rare in uncirculated. Although singles can be found, BU rolls are rarely encountered. Most of the issue was circulated – I once owned a 50 roll lot of circulated coins. Thus, at current prices this issue is underrated. As an uncirculated coin it is certainly as rare as the 39D.

The 42D is almost always a fully struck gem coin with at least five steps. Six step coins comprise a good percentage of the issue, but are scarce simply because the number of available BU coins is so low.

There is one major variety, exhibiting an upside down D mintmark under the normal mintmark. See variety section for details and photo.

Silver Alloy Wartime Nickels

1942P

Availability of Uncirculated Coins: singles available, rolls periodically available
Appearance of Uncirculated Coins: brilliant satin to frosty surfaces
Existence of Prooflike coins: rarely seen with semi or full prooflike surfaces
Five Step Frequency of Appearance: 1:10 Six Step Coin Rarity: very rare

The 42P in the past was always considered the rarest wartime nickel, however my experience leads me to believe that it is not that rare. The 42s seems to be just as scarce, and the 44D more scarce. However keep in mind that wartime nickels are very popular as a sub-type, and all of the issues command higher prices than their actual availability warrants. Large lots of this and all other wartime nickels exist.

1942P's are always very nice, without the surface roughness of the copper-nickel issue. Gems are easy to find.

1942S

Availability of Uncirculated Coins: singles available, rolls periodically available
Appearance of Uncirculated Coins: brilliant satin to frosty surfaces
Existence of Prooflike coins: periodically seen with semi or prooflike surfaces
Five Step Frequency of Appearance: 1:20 Six Step Coin Rarity: very rare

The 42S issue is usually very flashy, with surfaces tending toward prooflike.

In uncirculated, the 42S seems to have the same availability as the 42P. Since large lots of this and all other wartime nickels exist, the relative rarity could change as hoards come out.

There is a variety with a diebreak extending through the S, appearing as a $.

1943P

Availability of Uncirculated Coins: singles common, rolls are often available
Appearance of Uncirculated Coins: brilliant satin to frosty surfaces
Existence of Prooflike coins: rarely seen with semi or full prooflike surfaces
Five Step Frequency of Appearance: 1:5 Six Step Coin Rarity: scarce

One of the most common wartime nickels, probably due to its very high mintage. Some of the issue shows planchet roughness. Hoards do exist.

There are two major double die varieties for this date, the double eye and the 43/42 overdate. See variety section for details and photos.

1943D

Availability of Uncirculated Coins: singles common, rolls are often available
Appearance of Uncirculated Coins: brilliant satin almost prooflike surfaces
Existence of Prooflike coins: often seen with semi-prooflike surfaces
Five Step Frequency of Appearance: 1:2 Six Step Coin Rarity: common

Despite its relatively low mintage, the 43D, although in the past wildly promoted, is not really that rare, and hoards do exist.

This coin is almost always very flashy, exhibiting a full strike and very brilliant, almost semi-prooflike surfaces. True prooflike coins however are quite scarce. This issue is among the most common of the early Jeffersons in full five step condition. Many six-step coins are available, and a gem original roll may have as many as 20 or more six step coins.

1943S

Availability of Uncirculated Coins: singles common, rolls are often available
Appearance of Uncirculated Coins: brilliant satin to frosty to prooflike surfaces
Existence of Prooflike coins: often seen with semi to full prooflike surfaces
Five Step Frequency of Appearance: 1:10 Six Step Coin Rarity: scarce

Usually a very nice coin, though sometimes weakly struck. Of the S-mint wartime nickels, the 43S is the easiest to find with five or six full steps. Sometimes seen with dazzling prooflike surfaces, showing prominent die polish lines under magnification.

1944P

Availability of Uncirculated Coins: singles common, rolls are often available
Appearance of Uncirculated Coins: brilliant satin to frosty surfaces
Existence of Prooflike coins: rare with semi or full prooflike surfaces
Five Step Frequency of Appearance: 1:10 Six Step Coin Rarity: rare

This issue often exhibits surface roughness, however smooth coins can be found with some searching. Gem coins are somewhat difficult to find. As a roll, much harder to find than the 43P. Six step coins are seldom available.

1944D

Availability of Uncirculated Coins: singles usually available, rolls sometimes seen
Appearance of Uncirculated Coins: brilliant satin surfaces
Existence of Prooflike coins: true semi or prooflike coins are rare
Five Step Frequency of Appearance: 1:2 Six Step Coin Rarity: common

The 44D has similar characteristics to the 43D, although as an uncirculated coin, it is much scarcer.

This coin is almost always very nice, exhibiting a full strike and very brilliant, satiny surfaces. This issue is among the most common of the early Jeffersons in full five step condition. Many six-step coins are available, and a gem original roll may have as many as 20 or more six step coins.

1944S

Availability of Uncirculated Coins: singles usually available, rolls sometimes seen
Appearance of Uncirculated Coins: brilliant frosty surfaces
Existence of Prooflike coins: true semi or prooflike coins are very scarce
Five Step Frequency of Appearance: 1:100 Six Step Coin Rarity: very rare

The 44S does not come as nice as either 42S or 43S, and the coins are usually frosty, with some surface roughness evident. Much of the issue was struck from worn dies. Although rolls can be had, the 44S seems to be much scarcer than either the 43S or the 45S.

This is the second rarest wartime nickel with five steps, and six step coins are very rare.

Like the 1942S, some coins have a curious diebreak extending from the dome through the mintmark, called the dollar sign variety. In some cases, the diebreak extends into the step area, and such coins tend to have five or even six steps. Perfect die 44S coins with five full steps are quite rare, and six step coins are very rare – I have encountered only a few.

1945P

Availability of Uncirculated Coins: singles common, rolls sometimes available
Appearance of Uncirculated Coins: brilliant frosty surfaces
Existence of Prooflike coins: very rare with semi or full prooflike surfaces
Five Step Frequency of Appearance: 1:20 Six Step Coin Rarity: very rare

This issue usually exhibits surface roughness, and smooth gems are very difficult to find. As a roll, much harder to find than the 44P. Six step coins are very rare and seldom seen.

There are two major double die reverses collected for this date. See variety section for details and photos.

1945D

Availability of Uncirculated Coins: singles common, rolls sometimes available
Appearance of Uncirculated Coins: brilliant frosty surfaces
Existence of Prooflike coins: very rare with semi or full prooflike surfaces
Five Step Frequency of Appearance: 1:5 Six Step Coin Rarity: scarce

Unlike the other wartime D mints, the 45D is usually not that attractive. Normally the surfaces are rough, and the coin is normally not flashy or as well struck as earlier D mint coins.

1945S

Availability of Uncirculated Coins: singles common, rolls often available
Appearance of Uncirculated Coins: brilliant frosty surfaces
Existence of Prooflike coins: very rare with semi or full prooflike surfaces
Five Step Frequency of Appearance: 1:300 Six Step Coin Rarity: extra rare

The 45S is similar in appearance to the 44S, but generally worse! The coins are usually frosty, but are often seen with moderate to severe surface roughness. Much of the issue was struck from worn dies.

This date is the 'stopper' in assembling a set of five full step war nickels, as it is much rarer than the second toughest date, the 44S. Finding indisputable five step coins is a terrific challenge, since the steps are usually shallow and weakly struck. Many of what to appear to be five step coins have small bridges between the steps. Sharp, deep five step coins are rare. Six step coins are extra rare.

Copper Nickel Composition Resumed

1946

Availability of Uncirculated Coins: singles common, rolls sometimes available
Appearance of Uncirculated Coins: brilliant frosty surfaces
Existence of Prooflike coins: very rare with semi or full prooflike surfaces
Five Step Frequency of Appearance: 1:10 Six Step Coin Rarity: very scarce

A common issue, with large lots of uncirculated rolls extant. The usual appearance is granular surfaces but fully struck. Because of the normal rough appearance, most of the issue is less than choice. Five step coins are not very difficult to find, but the steps are often nicked and marked.

1946D

Availability of Uncirculated Coins: singles common, rolls sometimes available
Appearance of Uncirculated Coins: brilliant frosty surfaces
Existence of Prooflike coins: very rare with semi or full prooflike surfaces
Five Step Frequency of Appearance: 1:5 Six Step Coin Rarity: scarce

Although not as common as 1946, the 46D is also available in multiple roll quantities. Often the 46D has a slight orange peel effect in the fields and a full to medium strike.

One die variety has a D punched over an inverted D. See variety section.

1946S

Availability of Uncirculated Coins: singles common, rolls sometimes available
Appearance of Uncirculated Coins: brilliant smooth to prooflike surfaces
Existence of Prooflike coins: common with semi or full prooflike surfaces
Five Step Frequency of Appearance: 1:30   Six Step Coin Rarity: very scarce

Although the 46S is easily available in multiple roll quantities, the coins are often not well struck. Although finding five step coins with worn dies is easy, locating full strike five step coins can be a challenge.

1947

Availability of Uncirculated Coins: singles common, rolls sometimes available
Appearance of Uncirculated Coins: brilliant frosty surfaces
Existence of Prooflike coins: very rare with semi or full prooflike surfaces
Five Step Frequency of Appearance: 1:10   Six Step Coin Rarity: very rare

Much of this common issue is choice, although some of the dull color coins common to later P mint coins begins to show up with this issue. Most 1947's are well struck, and granular surfaces is not as often seen as on 1946 or 1948. The steps are often nicked and marked.

1947D

Availability of Uncirculated Coins: singles common, rolls sometimes available
Appearance of Uncirculated Coins: brilliant or subdued frosty surfaces
Existence of Prooflike coins: very rare with semi or full prooflike surfaces
Five Step Frequency of Appearance: 1:5   Six Step Coin Rarity: scarce

The 47D is a little scarcer than the 46D, and like the 46D is often seen with a slight orange peel effect in the fields and a full to medium strike.

1947S

Availability of Uncirculated Coins: singles common, rolls sometimes available
Appearance of Uncirculated Coins: brilliant to dull smooth to prooflike surfaces
Existence of Prooflike coins: often seen with semi or full prooflike surfaces
Five Step Frequency of Appearance: 1:30   Six Step Coin Rarity: very rare

Although sometimes seen with blazing luster, most coins seen have a grayish, even dirty appearance. True gem full strike coins are more difficult to locate than 46S or 48S or 49S.

1948

Availability of Uncirculated Coins: singles common, rolls sometimes available
Appearance of Uncirculated Coins: brilliant frosty surfaces
Existence of Prooflike coins: very rare with semi or full prooflike surfaces
Five Step Frequency of Appearance: 1:10   Six Step Coin Rarity: very scarce

Usually seen with granular surfaces.

1948D

Availability of Uncirculated Coins: singles common, rolls sometimes available
Appearance of Uncirculated Coins: brilliant or subdued frosty surfaces
Existence of Prooflike coins: extremely rare with semi or full prooflike surfaces
Five Step Frequency of Appearance: 1:10 Six Step Coin Rarity: scarce

This issue is available in multi-roll quantities and in the past has experienced wide price swings as a result of dealer promotions. Much of the issue is struck from worn dies and the coins exhibit flow lines, orange-peel and mushy lettering. Often seen with yellow or golden toning. Worn die five step coins are easy to find, however full strikes are difficult.

1948S

Availability of Uncirculated Coins: singles common, rolls sometimes available
Appearance of Uncirculated Coins: brilliant smooth to prooflike surfaces
Existence of Prooflike coins: often seen with semi or full prooflike surfaces
Five Step Frequency of Appearance: 1:20 Six Step Coin Rarity: rare

The 48S appears somewhat scarcer than the 46S or 47S. Often seen with blazing luster, generally the 48S is the nicest appearing S mint of the 46-54 run. They usually have a good strike and exhibit white blazing and often prooflike luster.

1949

Availability of Uncirculated Coins: singles common, rolls seldom available
Appearance of Uncirculated Coins: brilliant to subdued frosty surfaces
Existence of Prooflike coins: extremely rare with semi or full prooflike surfaces
Five Step Frequency of Appearance: 1:500 Six Step Coin Rarity: unverified

Unlike the earlier P mints, the 49 seems to be difficult to locate in multiple roll quantities. The 49 has the roughest looking surfaces of the 42-50 P mint run. Often has dull luster and weak strikes. The 49 is one of the most difficult of the P mints to locate in full five step condition; it's possible six step coins may not exist.

Combining the general poor appearance plus the actual rarity of full step coins gives us one of the rarest coins in the Jefferson series in gem, full strike, full step!

1949D

Availability of Uncirculated Coins: singles common, rolls seldom available
Appearance of Uncirculated Coins: brilliant frosty to subdued surfaces
Existence of Prooflike coins: extremely rare with semi or full prooflike surfaces
Five Step Frequency of Appearance: 1:20 Six Step Coin Rarity: rare

A scarcer D mint coin that often appears dull and spotty with soft strikes. Five step coins are harder to find than any of the 40 through 48 or the 50-52 D's. Six step coins are seldom seen.

One die has a D punched over an S. See variety section.

1949S

Availability of Uncirculated Coins: singles common, rolls sometimes available
Appearance of Uncirculated Coins: brilliant smooth to prooflike surfaces
Existence of Prooflike coins: often seen with semi or full prooflike surfaces
Five Step Frequency of Appearance: 1:150 Six Step Coin Rarity: extra rare

Somewhat difficult to locate in large quantities, although I once was able to purchase dozens of original rolls as one lot! Often weakly struck, however there are full strike blazers too.

1950

Availability of Uncirculated Coins: singles common, rolls sometimes available
Appearance of Uncirculated Coins: brilliant frosty to almost PL surfaces
Existence of Prooflike coins: scarce with semi or full prooflike surfaces
Five Step Frequency of Appearance: 1:150 Six Step Coin Rarity: very rare

Very underrated when compared to the 50D. Large quantities are not often available. The surfaces on 1950's range from very granular like earlier P mints, to very smooth almost prooflike surfaces. Fully semi prooflike coins are sometimes seen. The overall strike is generally good, but five step coins are tough to locate, though not as scarce as the 49.

1950D

Availability of Uncirculated Coins: singles common, rolls often available
Appearance of Uncirculated Coins: dull frosty surfaces
Existence of Prooflike coins: extremely rare with semi or full prooflike surfaces
Five Step Frequency of Appearance: 1:10 Six Step Coin Rarity: very scarce

Despite having the lowest mintage, the 50D is actually one of the most common early Jefferson in uncirculated, due to excessive hoarding. Large roll lots are periodically seen. Quality is an issue however, and the coins often have a dull luster and are struck from worn dies. Fully struck very brilliant coins are somewhat scarce. Full five step coins are easy to find.

There are 50D counterfeits made by the same folks who made the fake 39D's. The fake 50D's exhibit the same characteristics: a silvery gray appearance, and under magnification, the design details are not sharp, including the steps which are not sharp as they would be on a genuine 1950D. A diagnostic of this counterfeit is a raised projection to the right of the crossbar of the A in AMERICA. The counterfeits originated in the Midwest.

1951

Availability of Uncirculated Coins: singles common, rolls sometimes available
Appearance of Uncirculated Coins: brilliant to dark smooth to semi-PL surfaces
Existence of Prooflike coins: often seen with semi prooflike surfaces
Five Step Frequency of Appearance: 1:100 Six Step Coin Rarity: very rare

Rolls of 1951 nickels are not that difficult to buy, however most are neither brilliant nor gem. The 1951 begins to exhibit the characteristic of all of the early 1950's Philadelphia Mint coins – dull to dark color. The 51 is more available as a brilliant coin than the 52-55 nickels however, and unlike the later coins, many have a good strike. Brilliant five step coins are very difficult to locate.

1951D

Availability of Uncirculated Coins: singles common, rolls sometimes available
Appearance of Uncirculated Coins: subdued frosty surfaces
Existence of Prooflike coins: very rare with semi or full prooflike surfaces
Five Step Frequency of Appearance: 1:30 Six Step Coin Rarity: rare

Rolls are sometimes available, and the appearance ranges from brilliant to a grayish luster that is often uneven due to die wear. Much of the issue has a weak strike and many abrasions.

1951S

Availability of Uncirculated Coins: singles common, rolls sometimes available
Appearance of Uncirculated Coins: brilliant smooth to prooflike surfaces
Existence of Prooflike coins: often seen with semi or full prooflike surfaces
Five Step Frequency of Appearance: 1:150 Six Step Coin Rarity: extra rare

Large quantities of this date are usually difficult to locate. Most 51S exhibit a good strike for an S mint, but the luster varies from a gray look, similar to that seen on 1947S, to blazing prooflike. The 51S as a five full step coin is scarcer than the 49S, but not nearly as rare as 52S, 53S or 54S.

1952

Availability of Uncirculated Coins: singles common, rolls sometimes available
Appearance of Uncirculated Coins: subdued to dark smooth to semi-PL surfaces
Existence of Prooflike coins: sometimes seen with semi or full prooflike surfaces
Five Step Frequency of Appearance: 1:200 Six Step Coin Rarity: unverified

Though more common than the 1951, more of this issue is of very poor quality. Dark colored, poorly struck, nicked and marked coins abound. The 1952 is actually quite rare as a full strike, brilliant white coin, and is one of the rarest P mints in this condition with unmarked full steps.

1952D

Availability of Uncirculated Coins: singles common, rolls sometimes available
Appearance of Uncirculated Coins: brilliant frosty or smooth surfaces
Existence of Prooflike coins: rare with semi or full prooflike surfaces
Five Step Frequency of Appearance: 1:20 Six Step Coin Rarity: unverified

The 52D is usually brilliant and much nicer than the previous D mints, though often seen with scratches. Sometimes seen with soft hair by Jefferson's ear.

1952S

Availability of Uncirculated Coins: singles common, rolls often available
Appearance of Uncirculated Coins: brilliant smooth to semi or prooflike surfaces
Existence of Prooflike coins: often seen with semi or full prooflike surfaces
Five Step Frequency of Appearance: 1:300   Six Step Coin Rarity: unverified

As with all the later S mints, the 52S is often struck from dies exhibiting varying degrees of wear. That being said, it is easier to find full strike coins of this date than either 53S or 54S. Weak five full or even six full step coins are often encountered, however full strike full step coins are quite rare, and full strike six step coins are unverified.

This is not a rare coin by the roll, and quantities do exist.

1953

Availability of Uncirculated Coins: singles common, rolls often available
Appearance of Uncirculated Coins: subdued to dark smooth to semi-PL surfaces
Existence of Prooflike coins: often seen with semi or full prooflike surfaces
Five Step Frequency of Appearance: 1:500   Six Step Coin Rarity: unverified

Really a common coin if one cares nothing about quality. Most of this issue is of very poor quality. Dark colored, poorly struck, nicked and marked coins are the norm. The 1953 is rare as a full strike, brilliant white coin, and is one of the rarest P mints in full strike condition with unmarked full steps.

1953D

Availability of Uncirculated Coins: singles common, rolls usually available
Appearance of Uncirculated Coins: brilliant frosty surfaces
Existence of Prooflike coins: rare with semi or full prooflike surfaces
Five Step Frequency of Appearance: 1:200   Six Step Coin Rarity: unverified

This date is usually choice, with good luster and strike, although weakly struck hair by Jefferson's ear is often a problem. Gem coins with minimal marks are somewhat difficult to find. Gem coins, when found, usually have partial steps, and this issue is very difficult to locate as a full five step coin.

1953S

Availability of Uncirculated Coins: singles common, rolls usually available
Appearance of Uncirculated Coins: brilliant smooth to semi or prooflike surfaces
Existence of Prooflike coins: often seen with semi or full prooflike surfaces
Five Step Frequency of Appearance: 1:5000   Six Step Coin Rarity: unverified

Although easily found as a BU coin, most are struck from worn dies. Full strike blazers are periodically seen, but are rare. Most five step coins have a weak strike, and usually that is what the collector must settle for. Such coins are often touted as rare and offered at high prices. A full strike five full step coin is a major rarity, and few have been discovered. This is one of the real stoppers!

1954

Availability of Uncirculated Coins: singles common, rolls often available
Appearance of Uncirculated Coins: subdued to dark smooth to semi-PL surfaces
Existence of Prooflike coins: often seen with semi or full prooflike surfaces
Five Step Frequency of Appearance: 1:200 Six Step Coin Rarity: unverified

Most of this issue is of very poor quality. Dark colored, poorly struck, nicked and marked coins are commonly seen. The 1954 is rare as a full strike, brilliant white coin, however it is not nearly as rare in this condition as the 1952,1953 or 1955. The 1954 is quite scarce in full strike condition with unmarked full steps.

1954D

Availability of Uncirculated Coins: singles common, rolls often available
Appearance of Uncirculated Coins: brilliant frosty surfaces
Existence of Prooflike coins: scarce with semi or full prooflike surfaces
Five Step Frequency of Appearance: 1:200 Six Step Coin Rarity: unverified

This date is usually brilliant, with luster and strike that often leaves something to be desired! Weakly struck hair by Jefferson's ear is frequent. Gem coins with minimal marks are difficult to find. Gem coins, when found, usually have partial steps, and this issue is very difficult to locate as a full five step coin.

1954S

Availability of Uncirculated Coins: singles common, rolls often available
Appearance of Uncirculated Coins: brilliant smooth to semi or prooflike surfaces
Existence of Prooflike coins: often seen with semi or full prooflike surfaces
Five Step Frequency of Appearance: 1:3000 Six Step Coin Rarity: unverified

Dubbed "The King" by the former PAK Full Step Nickel Club because of its rarity as a fully struck full step coin, the 54S, though one of the top full step rarities, is by no means the rarest, being exceeded by the 53S, 60D, 61D, 68D and 69D.

Although easily found as a BU coin, most are struck from worn dies. Full strike blazers are periodically seen, but are rare. Most five step coins have a weak strike, and usually that is what the collector must settle for. Such coins are often touted as rare and offered at high prices. Full strike full step coins are a real find! There is one die variety with an S over D – see variety section.

1955

Availability of Uncirculated Coins: singles common, rolls often available
Appearance of Uncirculated Coins: subdued to dark smooth to semi-PL surfaces
Existence of Prooflike coins: often seen with semi or full prooflike surfaces
Five Step Frequency of Appearance: 1:200 Six Step Coin Rarity: unverified

Most of this issue is of "garbage" quality. Expect dark colored, poorly struck, nicked and marked coins. Very rare as a full strike, brilliant white coin. Although a common roll coin, in the past often promoted as rare.

1955D

Availability of Uncirculated Coins: singles common, rolls often available
Appearance of Uncirculated Coins: brilliant frosty , or smooth or granular surfaces
Existence of Prooflike coins: scarce with semi or full prooflike surfaces
Five Step Frequency of Appearance: 1:200 Six Step Coin Rarity: unverified

Although common in BU, as a full strike gem, the 55D is very scarce. Most have a rough appearance, and strike can be a problem. The date is scarce as a five step coin, and most seen are less than MS64.

There are a number of dies with a D over S mintmark. See variety section.

1956

Availability of Uncirculated Coins: singles common, rolls often available
Appearance of Uncirculated Coins: brilliant to blazing frosty surfaces
Existence of Prooflike coins: sometimes seen with semi prooflike surfaces
Five Step Frequency of Appearance: 1:10 Six Step Coin Rarity: rare

The 1956 is an enigma, as it is usually a stunningly beautiful coin in a P mint run of generally horrible coins. Fully struck gem full step coins are not a problem with this date

1956D

Availability of Uncirculated Coins: singles common, rolls often available
Appearance of Uncirculated Coins: brilliant frosty surfaces
Existence of Prooflike coins: scarce with semi or full prooflike surfaces
Five Step Frequency of Appearance: 1:150 Six Step Coin Rarity: unverified

Although always brilliant and not as granular as the 55D, the 56D is often weakly struck. Although weak strike full step coins are often seen, full strike gems are tough.

1957

Availability of Uncirculated Coins: singles common, rolls often available
Appearance of Uncirculated Coins: subdued to dark smooth to semi-PL surfaces
Existence of Prooflike coins: scarce with semi or full prooflike surfaces
Five Step Frequency of Appearance: 1:10 Six Step Coin Rarity: rare

Although often seen as a well struck somewhat bright coin, most of the issue is dark and poorly struck.

1957D

Availability of Uncirculated Coins: singles common, rolls often available
Appearance of Uncirculated Coins: brilliant frosty surfaces
Existence of Prooflike coins: scarce with semi or full prooflike surfaces
Five Step Frequency of Appearance: 1:100 Six Step Coin Rarity: rare

Most 1957D's are struck from worn dies, but full strike blazers are sometimes encountered – it is just a matter of finding the right batch of coins.

1958

Availability of Uncirculated Coins: singles common, rolls often available
Appearance of Uncirculated Coins: subdued to dark smooth to semi-PL surfaces
Existence of Prooflike coins: often seen with semi prooflike surfaces
Five Step Frequency of Appearance: 1:30 Six Step Coin Rarity: rare

Although the lowest mintage since 1955, this date is not rare. Quality is a problem, and much of the issue is dark or black in color. The surfaces usually have many small nicks and white, mark free gems are rare. Strike varies from full to very weak. Some 1958's are charcoal black and in the past these coins were promoted under the nickname "black beauties."

1958D

Availability of Uncirculated Coins: singles common, rolls often available
Appearance of Uncirculated Coins: brilliant frosty to smooth, semi-PL surfaces
Existence of Prooflike coins: sometimes seen with semi or full prooflike surfaces
Five Step Frequency of Appearance: 1:10 Six Step Coin Rarity: very scarce

Rolls of this date are either very nice and flashy, or very poor , although enough of the issue is nice that finding gem full strike coins is not too difficult. The 58D is an anomaly for late D-mints in that six step coins are available with surprising frequency.

1959

Availability of Uncirculated Coins: singles common, rolls often available
Appearance of Uncirculated Coins: subdued to dark smooth to semi-PL surfaces
Existence of Prooflike coins: often seen with semi prooflike surfaces
Five Step Frequency of Appearance: 1:10 Six Step Coin Rarity: rare

As with the preceding P-mints, quality is a problem, and much of the issue is dark or black in color with numerous surface nicks. White, mark free gems are scarce, but nowhere near as rare as 1958 or 1957. Strike varies from full to very weak. As with the 1958 issue, some 1959's appear as charcoal colored "black beauty" coins.

1959D

Availability of Uncirculated Coins: singles common, rolls often available
Appearance of Uncirculated Coins: brilliant frosty surfaces
Existence of Prooflike coins: rare with semi or full prooflike surfaces
Five Step Frequency of Appearance: 1:20 Six Step Coin Rarity: extra rare

The last reverse of 40 Denver mint coin that is available with decent surface condition, strike and steps.

1960

Availability of Uncirculated Coins: singles common, rolls often available
Appearance of Uncirculated Coins: brilliant frosty surfaces
Existence of Prooflike coins: rare with semi or full prooflike surfaces
Five Step Frequency of Appearance: 1:100 Six Step Coin Rarity: unverified

Although frosty, most coins show abrasions and bagmarks. Often the coins are struck from worn dies, with some wear so bad as to cause a doubled effect on the lettering and devices. Often Jefferson's hair by his ear is not fully struck up.

Note for 1960-67 issues

Many of the P and D mint issues of 1960 through 1967 show five steps at the right side of Monticello, but the steps will not connect to the left side of the stairs. This is unlike previous issues which almost always have connected steps at each side, the missing area being below the second pillar from the right. By now, the master hub is showing extreme wear, and the steps are usually shallow. Six step coins probably do not exist for most dates, except on proofs.

1960D

Availability of Uncirculated Coins: singles common, rolls often available
Appearance of Uncirculated Coins: brilliant frosty to semi prooflike surfaces
Existence of Prooflike coins: sometimes seen with semi or full prooflike surfaces
Five Step Frequency of Appearance: 1:5000 Six Step Coin Rarity: unknown

This issue is common without steps! Coins with even half the steps showing are scarce, and anything approaching five steps is rarely seen. As a fully struck five step coin, the 60D is one of the rarest coins in the set. Finding gem quality coins, even ignoring the steps, is a challenge. Frequently seen with soft hair by the ear, and showing excessive die wear.

1961

Availability of Uncirculated Coins: singles common, rolls often available
Appearance of Uncirculated Coins: brilliant frosty surfaces
Existence of Prooflike coins: rare with semi or full prooflike surfaces
Five Step Frequency of Appearance: 1:150 Six Step Coin Rarity: unknown

Although frosty, most coins show abrasions and bagmarks. Often the coins are struck from worn dies, with some wear so bad as to cause a doubled effect on the lettering and devices. Often Jefferson's hair by his ear is not fully struck up. The 1961 is by far the scarcest of the 1960-64 run in full five step condition.

1961D

Availability of Uncirculated Coins: singles common, rolls often available
Appearance of Uncirculated Coins: brilliant frosty to semi prooflike surfaces
Existence of Prooflike coins: sometimes seen with semi or full prooflike surfaces
Five Step Frequency of Appearance: 1:5000 Six Step Coin Rarity: unknown

Like the 60D, this issue is common without steps and coins with even half the steps showing are scarce. Anything approaching five steps is rare. As a fully struck five step coin, the 61D is one of the rarest coins in the set, ranking with the 60D. Gem quality coins are rare. Frequently seen with soft hair by the ear, and showing excessive die wear.

1962

Availability of Uncirculated Coins: singles common, rolls often available
Appearance of Uncirculated Coins: brilliant frosty surfaces
Existence of Prooflike coins: rare with semi or full prooflike surfaces
Five Step Frequency of Appearance: 1:20 Six Step Coin Rarity: unverified

Although frosty, most coins show abrasions and bagmarks. Often the coins are struck from worn dies, with some wear so bad as to cause a doubled effect on the lettering and devices. Often Jefferson's hair by his ear is not fully struck up. The general quality for 1962 is worse than for either 1960 or 1961.

1962D

Availability of Uncirculated Coins: singles common, rolls often available
Appearance of Uncirculated Coins: brilliant frosty to semi prooflike surfaces
Existence of Prooflike coins: sometimes seen with semi or full prooflike surfaces
Five Step Frequency of Appearance: 1:2000 Six Step Coin Rarity: unknown

This date is the worse of the 1960-1964 Denver mint coins. Most are very rough looking and poorly struck. Although the date is easier to find with five steps than the previous two years, it is still very rare in that condition. Gem coins are very difficult to find.

1963

Availability of Uncirculated Coins: singles common, rolls often available
Appearance of Uncirculated Coins: brilliant frosty surfaces
Existence of Prooflike coins: rare with semi or full prooflike surfaces
Five Step Frequency of Appearance: 1:20 Six Step Coin Rarity: unverified

Although frosty, most coins show abrasions and bagmarks, though not as frequently as the 1962. Often the coins are struck from worn dies, with some wear so bad as to cause a doubled effect on the lettering and devices. Often Jefferson's hair by his ear is not fully struck up. The general quality for 1963 is much better than for 1962.

1963D

Availability of Uncirculated Coins: singles common, rolls often available
Appearance of Uncirculated Coins: brilliant frosty to semi prooflike surfaces
Existence of Prooflike coins: sometimes seen with semi or full prooflike surfaces
Five Step Frequency of Appearance: 1:3000 Six Step Coin Rarity: unknown

The 1963D is similar to the 62D in terms of quality. Most are very rough looking and poorly struck. The 63D is rarer with five steps than the previous year. Gem coins are rare and seldom seen. As with the other D mints of 1960-1964, the 63D is rarely seen with more than three-quarters of the steps showing.

1964

Availability of Uncirculated Coins: singles common, rolls often available
Appearance of Uncirculated Coins: brilliant frosty surfaces
Existence of Prooflike coins: rare with semi or full prooflike surfaces
Five Step Frequency of Appearance: 1:30 Six Step Coin Rarity: unknown

With the very high mintage spread out over a couple years (due to the 1964 date freeze), the 1964 appears in all conditions from low to high quality, so finding a decent coin is possible with some patience.

1964D

Availability of Uncirculated Coins: singles common, rolls often available
Appearance of Uncirculated Coins: brilliant frosty to semi prooflike surfaces
Existence of Prooflike coins: sometimes seen with semi or full prooflike surfaces
Five Step Frequency of Appearance: 1:2000 Six Step Coin Rarity: unknown

As with the P mint of this year, a large number were produced over a couple years, and the quality varies. This date is often found with semi or prooflike surfaces, sometimes seen with a cameo effect. Five step coins are rare however, and the steps are always weak, due to the master hub wear.

1965

Availability of Uncirculated Coins: singles common, rolls sometimes available
Appearance of Uncirculated Coins: brilliant frosty surfaces
Existence of Prooflike coins: rare with semi or full prooflike surfaces
Five Step Frequency of Appearance: 1:100 Six Step Coin Rarity: unknown

The condition is similar to the 1964, however finding five step coins is a challenge. Most collectors give up quickly, and settle for a special mint set coin.

Rolls are much harder to find for this and subsequent dates, since the BU roll boom had faded out.

1966

Availability of Uncirculated Coins: singles common, rolls sometimes available
Appearance of Uncirculated Coins: brilliant frosty surfaces
Existence of Prooflike coins: rare with semi or full prooflike surfaces
Five Step Frequency of Appearance: 1:100 Six Step Coin Rarity: unknown

The same can be said here as for 1965, and most collectors use a SMS coin in their sets. For the first time, Felix Schlag's initials were added to the obverse.

1967

Availability of Uncirculated Coins: singles common, rolls sometimes available
Appearance of Uncirculated Coins: brilliant frosty surfaces
Existence of Prooflike coins: rare with semi or full prooflike surfaces
Five Step Frequency of Appearance: 1:500 Six Step Coin Rarity: unknown

The regular issue 1967 is the last coin produced with the original unmodified reverse of 1940, and by now the master hub was very worn. Most 1967's do not have very sharp detail, and five full step coins are quite rare, and will exhibit very shallow steps. Most collectors use a special mint set coin in their sets. It should be noted however, that the SMS coins used the modified reverse that was used on all subsequent issues through 1970.

Reverse modification of 1967

1968D

Availability of Uncirculated Coins: singles common, rolls sometimes available
Appearance of Uncirculated Coins: brilliant frosty to semi-prooflike surfaces
Existence of Prooflike coins: often seen with semi or full prooflike surfaces
Five Step Frequency of Appearance: 1:5000 Six Step Coin Rarity: unverified

The 68D is not as common as the earlier D mints, since few rolls were saved. In fact, many available rolls are assembled rolls, taken from mint sets that were cut up for the Kennedy Half.

The reverse modification did not result in an appreciable improvement in quality, and gem fully struck coins are difficult to find. The highest quality coins are found in the mint sets. As a full step coin, the 68D is one of the rarest coins in the series, and most 68D's have less than half the steps showing. A feature of the reverse modification is that part of the sixth step is usually showing.

1968S

Availability of Uncirculated Coins: singles common, rolls sometimes available
Appearance of Uncirculated Coins: brilliant frosty to semi-prooflike surfaces
Existence of Prooflike coins: often seen with semi or full prooflike surfaces
Five Step Frequency of Appearance: 1:500 Six Step Coin Rarity: rare

The first S mint nickel since 1954 caused a sensation at its issue, and there was some hoarding, thus the 68S is common as a BU coin. The quality description for the 68D also applies to the 68S, however the 68S is much easier to find in full five or even six step condition.

1969D

Availability of Uncirculated Coins: singles common, rolls sometimes available
Appearance of Uncirculated Coins: brilliant frosty to semi-prooflike surfaces
Existence of Prooflike coins: often seen with semi or full prooflike surfaces
Five Step Frequency of Appearance: 1:10000 Six Step Coin Rarity: unverified

Overall quality of the 69D is decent, however finding full strike MS65 coins can be challenging. As a full step coin, this date is the true king of the series. The 69D is incredibly rare as a five full step coin, and it is even very rare with more than half the steps showing. Since any five step 69D will have at least part of the sixth step showing, it is possible that a full six step coin may exist, but none have been reported.

1969S

Availability of Uncirculated Coins: singles common, rolls sometimes available
Appearance of Uncirculated Coins: brilliant frosty to semi-prooflike surfaces
Existence of Prooflike coins: often seen with semi or full prooflike surfaces
Five Step Frequency of Appearance: 1:2000 Six Step Coin Rarity: unverified

Similar in appearance and quality to the 69D, and overall a better coin than the 68S. With full steps however, the 69S is one of the rarer dates, thus it is collectible if more than 3/4 of the steps are showing.

1970D

Availability of Uncirculated Coins: singles common, rolls sometimes available
Appearance of Uncirculated Coins: brilliant frosty to semi-prooflike surfaces
Existence of Prooflike coins: often seen with semi or full prooflike surfaces
Five Step Frequency of Appearance: 1:3000 Six Step Coin Rarity: unverified

Another major full step rarity, although not nearly as rare as the 68D or 69D. As with any of the rarest full step dates, the best hope may be to acquire an "almost five step" coin.

Because the 70D Kennedy Half was available only in the mint sets, a huge number of 1970 mint sets were cut up for the half, and thus most 70D nickel rolls are assembled rolls taken from the mint sets.

1970S

Availability of Uncirculated Coins: singles common, rolls sometimes available
Appearance of Uncirculated Coins: brilliant frosty to semi-prooflike surfaces
Existence of Prooflike coins: often seen with semi or full prooflike surfaces
Five Step Frequency of Appearance: 1:100 Six Step Coin Rarity: very scarce

This is the easiest full step coin to find in the 1968-1970 date run, but the challenge is to find mark free coins, as many exhibit scratches and abrasions. The 70S can be found with six steps with some diligent searching.

Reverse of 1971

Overall characteristics:

The reverse of 1971, in use through 1976, resulted is some of the best looking nickels in a long time. The coins are generally very well struck, and full step coins are relatively easy to find, although they are progressively scarcer as the years go on, probably due to master hub wear. The sixth step if present is always very thin.

Original BU rolls are generally hard to find, and when found the coins will exhibit a more satin appearance. Most rolls are assembled from cut-up mint sets, and the coins are usually very flashy often with semi or prooflike appearance. However, the coins high luster makes them more susceptible to abrasions, and true gem coins from mint sets are for the most part scarce. The reverse of 71 nickels are often golden toned.

1971

Availability of Uncirculated Coins: singles common, rolls seldom available
Appearance of Uncirculated Coins: brilliant, satin to flashy and blazing
Existence of Prooflike coins: often seen with semi prooflike surfaces
Five Step Frequency of Appearance: 1:3 Six Step Coin Rarity: scarce

1971D

Availability of Uncirculated Coins: singles common, rolls seldom available
Appearance of Uncirculated Coins: brilliant, satin to flashy and blazing
Existence of Prooflike coins: often seen with semi prooflike surfaces
Five Step Frequency of Appearance: 1:2 Six Step Coin Rarity: scarce

1972

Availability of Uncirculated Coins: singles common, rolls seldom available
Appearance of Uncirculated Coins: brilliant, satin to flashy and blazing
Existence of Prooflike coins: often seen with semi prooflike surfaces
Five Step Frequency of Appearance: 1:5 Six Step Coin Rarity: scarce

1972D

Availability of Uncirculated Coins: singles common, rolls seldom available
Appearance of Uncirculated Coins: brilliant, satin to flashy and blazing
Existence of Prooflike coins: often seen with semi prooflike surfaces
Five Step Frequency of Appearance: 1:10 Six Step Coin Rarity: very scarce

1973

Availability of Uncirculated Coins: singles common, rolls seldom available
Appearance of Uncirculated Coins: brilliant, satin to flashy and blazing
Existence of Prooflike coins: often seen with semi prooflike surfaces
Five Step Frequency of Appearance: 1:3 Six Step Coin Rarity: scarce

1973D

Availability of Uncirculated Coins: singles common, rolls seldom available
Appearance of Uncirculated Coins: brilliant, satin to flashy and blazing
Existence of Prooflike coins: often seen with semi prooflike surfaces
Five Step Frequency of Appearance: 1:3 Six Step Coin Rarity: scarce

1974

Availability of Uncirculated Coins: singles common, rolls seldom available
Appearance of Uncirculated Coins: brilliant, satin to flashy and blazing
Existence of Prooflike coins: often seen with semi prooflike surfaces
Five Step Frequency of Appearance: 1:10 Six Step Coin Rarity: very scarce

1974D

Availability of Uncirculated Coins: singles common, rolls seldom available
Appearance of Uncirculated Coins: brilliant, satin to flashy and blazing
Existence of Prooflike coins: often seen with semi prooflike surfaces
Five Step Frequency of Appearance: 1:10 Six Step Coin Rarity: very scarce

1975

Availability of Uncirculated Coins: singles common, rolls seldom available
Appearance of Uncirculated Coins: brilliant, satin to flashy and blazing
Existence of Prooflike coins: often seen with semi prooflike surfaces
Five Step Frequency of Appearance: 1:30 Six Step Coin Rarity: very scarce

1975D

Availability of Uncirculated Coins: singles common, rolls seldom available
Appearance of Uncirculated Coins: brilliant, satin to flashy and blazing
Existence of Prooflike coins: often seen with semi prooflike surfaces
Five Step Frequency of Appearance: 1:30 Six Step Coin Rarity: very scarce

1976

Availability of Uncirculated Coins: singles common, rolls seldom available
Appearance of Uncirculated Coins: brilliant, satin to flashy and blazing
Existence of Prooflike coins: often seen with semi prooflike surfaces
Five Step Frequency of Appearance: 1:1000 Six Step Coin Rarity: very rare

The 1976, is much rarer with five full steps than the other dates with the reverse of 1971. No doubt some of the cause is the wearing of the master hub, and perhaps mint personnel were too preoccupied with the bicentennial coins to worry about nickels! The color is often a subdued silver-gray.

1976D

Availability of Uncirculated Coins: singles common, rolls seldom available
Appearance of Uncirculated Coins: brilliant, satin to flashy and blazing
Existence of Prooflike coins: often seen with semi prooflike surfaces
Five Step Frequency of Appearance: 1:100 Six Step Coin Rarity: rare

Reverse of 1977

Overall characteristics:

The reverse of 1977 was used into 1982. The coins are generally very well struck until 1980, when die wear becomes a problem. After 1979, many of the coins show excessive wear on the lettering, evidence of erosion in the fields and orange-peel effect. Finding full step coins is a spotty experience, the frequency of appearance varying from year to year. The sixth step is thicker than on the previous reverse design.

Original BU rolls are generally hard to find, with the coins exhibiting a satin to frosty appearance. The 1977-1979 issues often are a subdued silver-gray color. Many rolls were assembled from cut-up mint sets, and those coins are more flashy but also more abraded and scratched.

1977

Availability of Uncirculated Coins: singles common, rolls seldom available
Appearance of Uncirculated Coins: brilliant, satin to flashy and blazing
Existence of Prooflike coins: sometimes seen with semi prooflike surfaces
Five Step Frequency of Appearance: 1:200 Six Step Coin Rarity: very rare

1977D

Availability of Uncirculated Coins: singles common, rolls seldom available
Appearance of Uncirculated Coins: brilliant, satin to flashy and blazing
Existence of Prooflike coins: sometimes seen with semi prooflike surfaces
Five Step Frequency of Appearance: 1:10 Six Step Coin Rarity: very scarce

1978

Availability of Uncirculated Coins: singles common, rolls seldom available
Appearance of Uncirculated Coins: brilliant, satin to flashy and blazing
Existence of Prooflike coins: sometimes seen with semi prooflike surfaces
Five Step Frequency of Appearance: 1:30 Six Step Coin Rarity: extra rare

1978D

Availability of Uncirculated Coins: singles common, rolls seldom available
Appearance of Uncirculated Coins: brilliant, satin to flashy and blazing
Existence of Prooflike coins: sometimes seen with semi prooflike surfaces
Five Step Frequency of Appearance: 1:10 Six Step Coin Rarity: very scarce

1979

Availability of Uncirculated Coins: singles common, rolls seldom available
Appearance of Uncirculated Coins: brilliant, satin to flashy and blazing
Existence of Prooflike coins: sometimes seen with semi prooflike surfaces
Five Step Frequency of Appearance: 1:150 Six Step Coin Rarity: extra rare

1979D

Availability of Uncirculated Coins: singles common, rolls seldom available
Appearance of Uncirculated Coins: brilliant, satin to flashy and blazing
Existence of Prooflike coins: sometimes seen with semi prooflike surfaces
Five Step Frequency of Appearance: 1:10 Six Step Coin Rarity: very scarce

1980P

Availability of Uncirculated Coins: singles common, rolls seldom available
Appearance of Uncirculated Coins: brilliant, satin to flashy and blazing
Existence of Prooflike coins: sometimes seen with semi prooflike surfaces
Five Step Frequency of Appearance: 1:20 Six Step Coin Rarity: very rare

This date and subsequent issues using this reverse are often seen showing die wear that can be extreme. Thus finding sharply struck coins with good , smooth surfaces can present a challenge.

1980D

Availability of Uncirculated Coins: singles common, rolls seldom available
Appearance of Uncirculated Coins: brilliant, satin to flashy and blazing
Existence of Prooflike coins: sometimes seen with semi prooflike surfaces
Five Step Frequency of Appearance: 1:10 Six Step Coin Rarity: very rare

1981P

Availability of Uncirculated Coins: singles common, rolls seldom available
Appearance of Uncirculated Coins: brilliant, satin to flashy and blazing
Existence of Prooflike coins: sometimes seen with semi prooflike surfaces
Five Step Frequency of Appearance: 1:30 Six Step Coin Rarity: unverified

1981D

Availability of Uncirculated Coins: singles common, rolls seldom available
Appearance of Uncirculated Coins: brilliant, satin to flashy and blazing
Existence of Prooflike coins: sometimes seen with semi prooflike surfaces
Five Step Frequency of Appearance: 1:20 Six Step Coin Rarity: unverified

1982P

Availability of Uncirculated Coins: singles common, rolls seldom available
Appearance of Uncirculated Coins: brilliant, satin to flashy and blazing
Existence of Prooflike coins: sometimes seen with semi prooflike surfaces
Five Step Frequency of Appearance: 1:50 Six Step Coin Rarity: unverified

1982D

Availability of Uncirculated Coins: singles common, rolls seldom available
Appearance of Uncirculated Coins: brilliant, satin to flashy and blazing
Existence of Prooflike coins: sometimes seen with semi prooflike surfaces
Five Step Frequency of Appearance: 1:30 Six Step Coin Rarity: unverified

Reverse of 1982

Overall characteristics:

The reverse of 1982 was used through 1986. The coins vary from very well struck to excessively worn, the die wear problem remaining unsolved. Many of the coins show excessive wear on the lettering, evidence of erosion in the fields and orange-peel effect. Full step coins are relatively easy to find, although they are progressively scarcer as the years go on, probably due to master hub wear. The Mint began to use left over proof reverse dies in Denver, and semi to prooflike reverses are often seen on Denver Mint issues.

Original BU rolls are generally hard to find. Some rolls were assembled from cut-up mint sets, when such sets were available, but the practice dropped off sharply after 1984 as mint sets became more expensive. The mint set coins are more flashy but also more abraded and scratched.

1982P

Availability of Uncirculated Coins: singles common, rolls seldom available
Appearance of Uncirculated Coins: brilliant, satin to frosty to flashy and blazing
Existence of Prooflike coins: sometimes seen with semi prooflike surfaces
Five Step Frequency of Appearance: 1:5 Six Step Coin Rarity: scarce

1982D

Availability of Uncirculated Coins: singles common, rolls seldom available
Appearance of Uncirculated Coins: brilliant, satin to frosty to flashy and blazing
Existence of Prooflike coins: sometimes seen with semi prooflike surfaces
Five Step Frequency of Appearance: 1:5 Six Step Coin Rarity: scarce

1983P

Availability of Uncirculated Coins: singles common, rolls seldom available
Appearance of Uncirculated Coins: brilliant, satin to frosty to flashy and blazing
Existence of Prooflike coins: sometimes seen with semi prooflike surfaces
Five Step Frequency of Appearance: 1:20 Six Step Coin Rarity: very scarce

1983D

Availability of Uncirculated Coins: singles common, rolls seldom available
Appearance of Uncirculated Coins: brilliant, satin to frosty to flashy and blazing
Existence of Prooflike coins: sometimes seen with semi prooflike surfaces
Five Step Frequency of Appearance: 1:20 Six Step Coin Rarity: very scarce

1984P

Availability of Uncirculated Coins: singles common, rolls seldom available
Appearance of Uncirculated Coins: brilliant, satin to frosty to flashy and blazing
Existence of Prooflike coins: sometimes seen with semi prooflike surfaces
Five Step Frequency of Appearance: 1:30 Six Step Coin Rarity: rare

1984D
Availability of Uncirculated Coins: singles common, rolls seldom available
Appearance of Uncirculated Coins: brilliant, satin to frosty to flashy and blazing
Existence of Prooflike coins: sometimes seen with semi prooflike surfaces
Five Step Frequency of Appearance: 1:30 Six Step Coin Rarity: rare

1985P
Availability of Uncirculated Coins: singles common, rolls seldom available
Appearance of Uncirculated Coins: brilliant, satin to frosty to flashy and blazing
Existence of Prooflike coins: sometimes seen with semi prooflike surfaces
Five Step Frequency of Appearance: 1:40 Six Step Coin Rarity: very rare

1985D
Availability of Uncirculated Coins: singles common, rolls seldom available
Appearance of Uncirculated Coins: brilliant, satin to frosty to flashy and blazing
Existence of Prooflike coins: sometimes seen with semi prooflike surfaces
Five Step Frequency of Appearance: 1:40 Six Step Coin Rarity: very rare

1986P
Availability of Uncirculated Coins: singles common, rolls seldom available
Appearance of Uncirculated Coins: brilliant, satin to frosty to flashy and blazing
Existence of Prooflike coins: sometimes seen with semi prooflike surfaces
Five Step Frequency of Appearance: 1:40 Six Step Coin Rarity: very rare

1986D
Availability of Uncirculated Coins: singles common, rolls seldom available
Appearance of Uncirculated Coins: brilliant, satin to frosty to flashy and blazing
Existence of Prooflike coins: sometimes seen with semi prooflike surfaces
Five Step Frequency of Appearance: 1:40 Six Step Coin Rarity: very rare

Reverse of 1987 and later changes

The changes to the master hub in 1987 resulted in overall sharper nickels, with generally much better steps. Most nickels dated 1987 and later have at least five full steps, and the coins are generally flashy. There is a problem with abrasions and scratches, so finding mark free gems can be a challenge. The die wear problems of the 80's seem to lessen considerably, with excessively worn dies less frequent, especially after 1989. As with previous modern issues, the flashiest coins are to be found in the mint sets.

1987P

Availability of Uncirculated Coins: singles common, rolls commonly available
Appearance of Uncirculated Coins: brilliant, satin to frosty to flashy and blazing
Existence of Prooflike coins: sometimes seen with semi prooflike surfaces
Five Step Frequency of Appearance: 1:2 Six Step Coin Rarity: scarce

1987D

Availability of Uncirculated Coins: singles common, rolls commonly available
Appearance of Uncirculated Coins: brilliant, satin to frosty to flashy and blazing
Existence of Prooflike coins: sometimes seen with semi prooflike surfaces
Five Step Frequency of Appearance: 1:2 Six Step Coin Rarity: scarce

1988P

Availability of Uncirculated Coins: singles common, rolls commonly available
Appearance of Uncirculated Coins: brilliant, satin to frosty to flashy and blazing
Existence of Prooflike coins: sometimes seen with semi prooflike surfaces
Five Step Frequency of Appearance: 1:3 Six Step Coin Rarity: scarce

1988D

Availability of Uncirculated Coins: singles common, rolls commonly available
Appearance of Uncirculated Coins: brilliant, satin to frosty to flashy and blazing
Existence of Prooflike coins: sometimes seen with semi prooflike surfaces
Five Step Frequency of Appearance: 1:3 Six Step Coin Rarity: scarce

1989P

Availability of Uncirculated Coins: singles common, rolls commonly available
Appearance of Uncirculated Coins: brilliant, satin to frosty to flashy and blazing
Existence of Prooflike coins: sometimes seen with semi prooflike surfaces
Five Step Frequency of Appearance: 1:5 Six Step Coin Rarity: common

1989D

Availability of Uncirculated Coins: singles common, rolls commonly available
Appearance of Uncirculated Coins: brilliant, satin to frosty to flashy and blazing
Existence of Prooflike coins: sometimes seen with semi prooflike surfaces
Five Step Frequency of Appearance: 1:5 Six Step Coin Rarity: common

1990P

Availability of Uncirculated Coins: singles common, rolls commonly available
Appearance of Uncirculated Coins: brilliant, satin to frosty to flashy and blazing
Existence of Prooflike coins: sometimes seen with semi prooflike surfaces
Five Step Frequency of Appearance: 1:5   Six Step Coin Rarity: common

1990D

Availability of Uncirculated Coins: singles common, rolls commonly available
Appearance of Uncirculated Coins: brilliant, satin to frosty to flashy and blazing
Existence of Prooflike coins: sometimes seen with semi prooflike surfaces
Five Step Frequency of Appearance: 1:5   Six Step Coin Rarity: common

1991P through 2000P

Availability of Uncirculated Coins: singles common, rolls commonly available
Appearance of Uncirculated Coins: brilliant, satin to frosty to flashy and blazing
Existence of Prooflike coins: sometimes seen with semi prooflike surfaces
Five Step Frequency of Appearance: 1:2   Six Step Coin Rarity: common

1991D through 2000D

Availability of Uncirculated Coins: singles common, rolls commonly available
Appearance of Uncirculated Coins: brilliant, satin to frosty to flashy and blazing
Existence of Prooflike coins: sometimes seen with semi prooflike surfaces
Five Step Frequency of Appearance: 1:2   Six Step Coin Rarity: common

During the 1990's, the Mint has periodically tweaked the master hubs, with the result that all of the 1990's coins are sharp with at least five steps – in fact, a 1990's Jefferson Nickel without full steps is a rarity. The flashy luster does make the coins prone to showing abrasions, and one must search through a number of coins to find a gem coin.

As of 2002, the description above still holds true.

Proof Coins

Proof Jefferson Nickels have been issued for all years except 1943-1949 and 1965-1967. Surface preparation follows the general Mint practice at time of issue. Brilliant proofs were issued from 1938-1942. During and after World War II production of proof coins was suspended.

When proof production was resumed in 1950, some of the early production exhibits an almost satin appearance in the fields. Later 1950 production was restored to full mirrored fields.

Proofs were again suspended for the dates 1965-1967. The Mint did issue Special Mint Sets containing coins that were struck once from polished dies. These SMS coins were progressively more like proofs as each year was produced, and many of the 1967 coins are as mirror-like, though not as well struck, as proofs.

Proof set production resumed in 1968, with production responsibility transferred to San Francisco and has continued to date. The proofs after 1970 exhibit a cameo contrast as part of the production process.

During the 1990's, the Mint became very aggressive in marketing their products to collectors, with the result that some special finish proof nickels were produced . A Thomas Jefferson Coin and Currency Set was produced in 1994 that contained a Matte Proof Nickel. A similar approach was taken in 1997 when the U.S. Botanic Gardens Coin and Currency Set was issued with a matte finish nickel.

1938

The first year of issue has the Reverse of 1938 design, and the step area is not well defined even on the proofs. Most 1938 proofs do not have a full sixth step all the way across. When all of step six is present, it will be thin at the third pillar.

Although very rare, some cameo frosted proofs exist. None exhibit the degree of frosting that post 1950 proofs sometimes show.

One hundred fifty 1938 proofs were encased in picture frames in 1939, each one being numbered and signed by Felix Schlag. In the 1960's Detroit coin dealer Earl Schill obtained a hoard of these encased proofs and sold them for $225.00 each. Since then, these encased proofs show up sporadically at auctions. They typically sold for about $800.00 in the 1990's, although serial number 1, said to have been owned by Schlag himself, sold for $5060.00 in January 1990 at Bowers and Merena's R.H. Gore Jr. And Dennis Irving Long auction sale. This particular encased proof had sold in November 1987 for $3410.00 at Bowers and Merena's Ebenezer Milton Saunders sale.

A JEFFERSON TRIBUTE

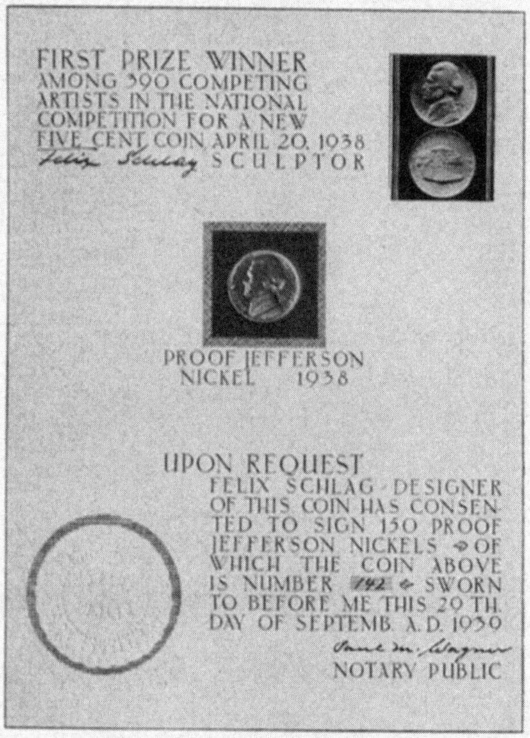

1938 PROOF JEFFERSON NICKEL

AUTOGRAPHED PRESENTATION PIECE

1938 Brilliant Proof Five Cent mounted on a strikingly beautiful presentation card (approximately 8 x 11). Legend on "Page" as follows:

"First prize winner among 390 competing artists in the National Competition for a New Five Cent Coin, April 20, 1938, Felix Schlag, sculptor."

Small cut showing obverse and reverse of rejected design. Mounted in the center is the prize winning proof nickel. Below is the undisputed seal and testimony of Paul Wagner, Notary Public.

Unique and rare, this showpiece has a guarantee of 150 only and each one numbered. Positively no more can exist. Seen at auction, one sold for a record $300.00.

The advanced collector knows this original rates exhibition and praise at home or in the office. YOURS FOR $225.00.

Shipped Insured Prepaid — Satisfaction Guaranteed

BE SURE! KNOW YOUR DEALER!

EARL C. SCHILL

Established 1927 . . . Your Downtown Dealer

1434 Farmer St. **Detroit, Michigan 48226**

P.N.G. Member No. 55 **A.N.A. No. 5700** **L.M. No. 175**

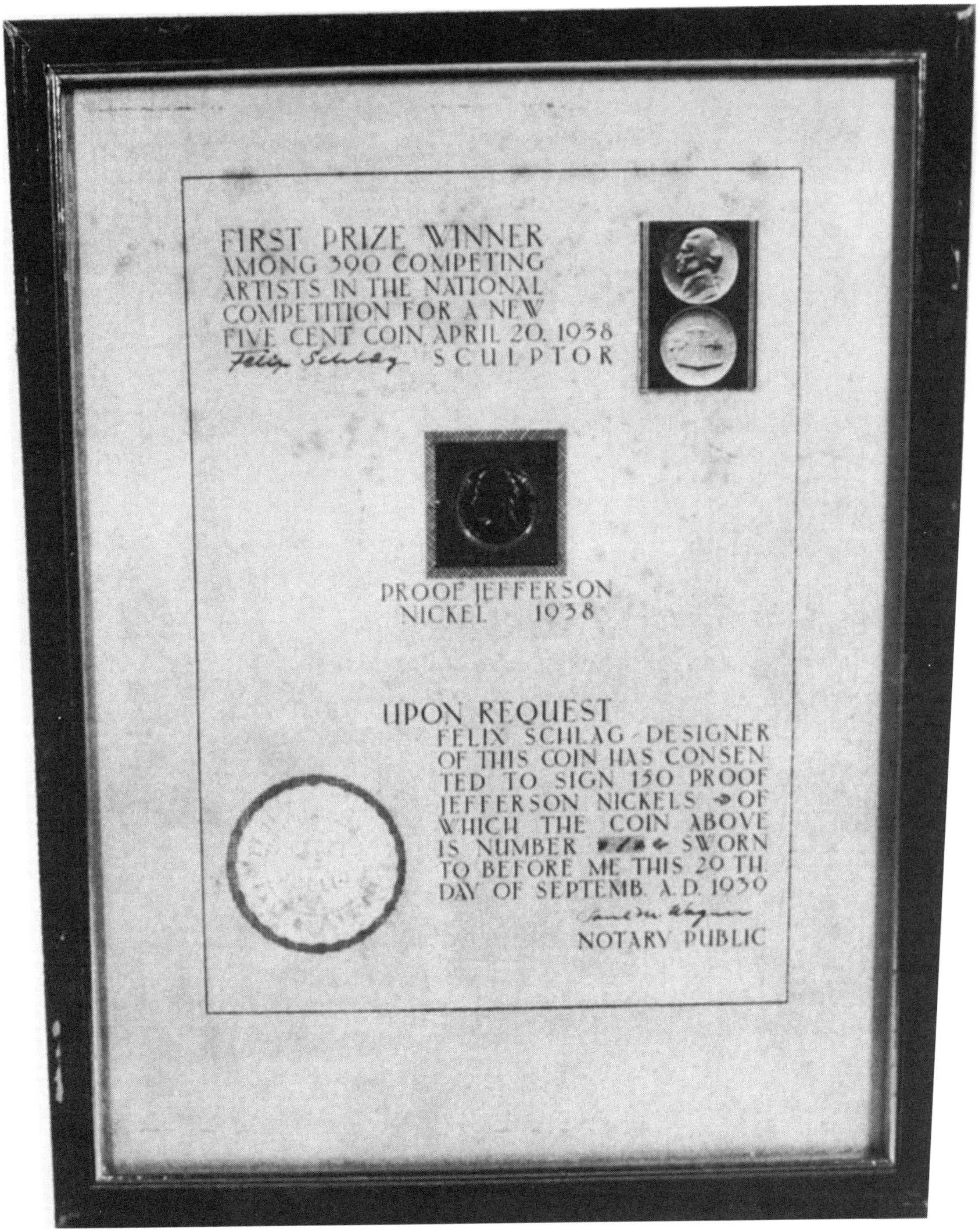
FIRST PRIZE WINNER
AMONG 390 COMPETING
ARTISTS IN THE NATIONAL
COMPETITION FOR A NEW
FIVE CENT COIN APRIL 20, 1938
SCULPTOR
PROOF JEFFERSON
NICKEL 1938
UPON REQUEST
FELIX SCHLAG DESIGNER
OF THIS COIN HAS CONSEN
TED TO SIGN 150 PROOF
JEFFERSON NICKELS OF
WHICH THE COIN ABOVE
IS NUMBER SWORN
TO BEFORE ME THIS 20 TH
DAY OF SEPTEMB. A.D. 1939
NOTARY PUBLIC

1939

Like the general circulation issue, the proofs were produced with both reverse types. Most of the proof issue has the Reverse of 1938, and like the 1938 proofs, the steps are not well defined. Six step proofs are very scarce. Cameos are rare.

A small part of the proof issue exhibits the Reverse of 1940 design with very sharp steps. All that I have seen have a full six steps.

The existence of 1939 proofs with the Reverse of 1940 design was not publicized until error specialist Tom Miller found one at a coin show in the late 1970's. The variety was subsequently touted by the PAK Full Step Nickel Club. Thirty of the coins were dispersed by Bowers and Merena at auction in 1985 as part of the Abe Kosoff estate. Otherwise, the Reverse of 1940 proof is seldom available.

1940

One proof die with the Reverse of 1938 somehow slipped into production in 1940. This variety was discovered by numismatist Bill Fivaz at the same coin show that Tom Miller had found the 1939 proof with the Reverse of 1940!

The 1940 proof with the Reverse of 1938 is rarely encountered, and most have only 5 1/2 steps at best. The coin may not exist with 6 steps. For some reason, many of this variety have tiny carbon spots.

Most of the 1940 issue has the sharp-stepped Reverse of 1940, and they always have six steps. Cameo proofs are rare.

1941-1942P

The only problem encountered is that some coins have an abrasion like area on Jefferson's shoulder. Otherwise, every date is well struck, with six steps. As with all of the pre-1950 proofs, cameo frosted proofs are rare. If found, the contrast is subdued.

1950-1964

As previously noted, some of the 1950 issue exhibits a satin like proof field. Most of the years have six steps, although the 1953 date is often seen with less than six steps. The steps are progressively weaker as the years progress due to master hub wear. By the 1960's, the sixth step is literally wearing thin! Cameo contrast proofs are available for all dates, with the earlier years being scarcer and exhibiting less contrast than the later years.

There is a collectible variety in 1955, a triple die reverse. The variety shows the doubling most prominently on the O of MONTICELLO. Most of these have been discovered in the "flat pack" packaging for the 1955 proof set.

1965-1967

No proofs were struck, however the Special Mint Set coins are frequently included in the proof series. Six step coins, as well as cameo frosted coins can be hard to locate. The 1965 coins are less mirror-like than the 1966 or 1967 issues.

The 1967 SMS nickel uses the modified Reverse of 1940 that was used on subsequent issues through 1970. See p. 19 regarding the so-called 1966 proofs.

1968S-1981S

Quality and strike is high for this entire run. The mintmark punch was replaced in 1979 and 1981 resulting in two mintmark varieties for those years. In both cases, the new mintmark is scarcer, particularly for the 1981.

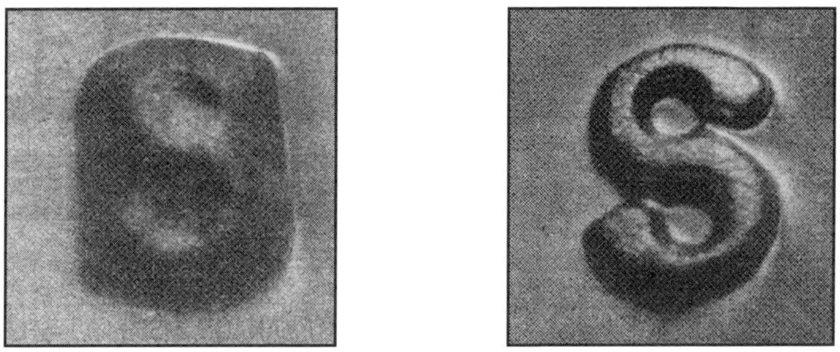

S mintmarks for 1979 proofs. At left is the old mintmark, appearing as a blob. At right is the new mintmark, always appearing as sharp and clear.

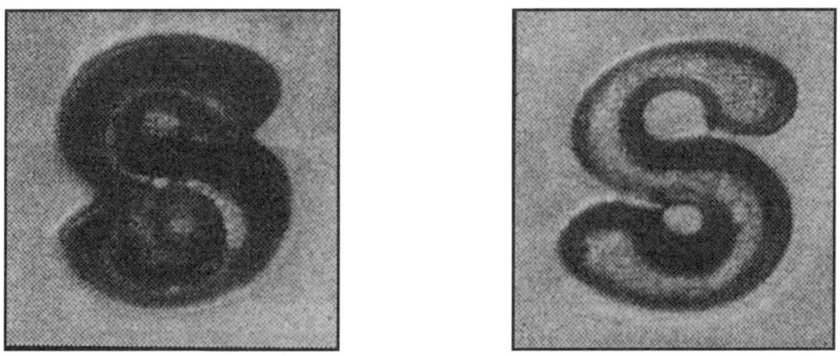

S mintmarks for 1981 proofs. At left is the old 1979 mintmark. At right is the new mintmark. Sharpness may vary, depending upon how deeply the S was punched.

1982S

The 1982S nickels, like the 1982P and D circulation strikes, are available with both obverse and reverse types. Hardly anyone pays any attention to this fact, and the differentiation is seldom noted. The new reverse is somewhat more common than the old. As with the circulation strikes, the steps are sharper on the new reverse.

1982S proof nickel with old Reverse of 1977. Note lack of detail.

1982S proof nickel with new Reverse of 1982. Note sharp detail.

1983S through 2000S

There is nothing to say about the mass produced, mundane issues after 1982! They are all spectacular in appearance, and all have sharp steps.

Interestingly, all of the later Jefferson proofs tend to tone to iridescent hues.

The special Matte Proof issues of 1994 and 1997 were discussed at the beginning of this section.

VARIETIES

This section is concerned with the major collectible varieties of the Jefferson Nickel Series. Before the late 1970's, the 1954S/D and the 1955D/S were the only varieties usually collected by most Jefferson Nickel collectors. The PAK Full Step Nickel Club publicized and promoted a number of other varieties, and many of them were later listed in 'The Red Book" and recognized by grading services.

1939 Double Die Reverse

There are two major double die reverses for this year, and a number of minor ones. All of them have the reverse of 1940. The major Variety I is listed in "The Red Book" and has been known almost since its release. It was discovered in 1939 by Malcolm O. E. Chell-Frost of Boston. The variety was very rare in uncirculated until a hoard was disbursed by Bowers and Merena after 1988. Some of the hoard coins were lightly carbon spotted.

The major variety II was not generally known until PAK promoted it. It is much rarer in uncirculated than the variety I

Both varieties are similar to the 1939 in terms of full step rarity, having a five step frequency of appearance of 1: 10. Six step coins may exist, however none have been reported.

1939 Double Die Reverse Variety I

1939 Double Die Reverse Variety II

1941S Large S

The San Francisco Mint replaced the mintmark punch in 1941, with the result that nickels made that year have both the old and new type. The new S is larger than the old, and less than 10 dies apparently had the new mintmark, thus this variety is very rare in uncirculated. The large S coins are usually well struck. Despite a five step frequency of appearance of 1:10, five step coins are scarce since the issue itself is scarce. No six step coins have been reported, however it's possible that some may exist.

1941Large S

1942D over horizontal D

This variety is possibly the rarest major Jefferson Variety in uncirculated. A small number of the coins were discovered in the late 1970's and disbursed. Few have come to light since. Since the 1942D nickel itself is rarely seen by the roll, the variety may remain very rare unless an unusual number show up is an unsearched roll.

Virtually all uncirculated examples show at least five steps, and like the 42D are extremely nice, full strike coins. The five step frequency of appearance is 1: 5. Six step coins probably exist.

1942D / Horizontal D

1942S (mintmark in normal location)

The first edition of this book erroneously listed this unique coin as a copper-nickel issue. I was able to purchase the exact coin in the 1980's, and it turns out that the coin was struck on what appears to be a silver wartime planchet. The coin is heavily circulated, and there is a lamination defect present. The ANA Certification Service was unable to verify that the coin was in fact the silver wartime composition, their tests showing a higher percent silver on the coin's surface than would be warranted. This could be due to circulation wear affecting the surface composition, or its possible the planchet is of some experimental composition.

The coin has the large S mintmark introduced in 1941 in the normal location to the right of Monticello.

The history of this coin is a mystery. It was found in circulation in 1961 by a collector, Ken Frith. The coin sold in 1974 for $11,000. at Bowers and Ruddy's Stanislaw Herstal Sale. The coin was auctioned again in 1975 at the Suburban Washington Sale. The coin was subsequently owned by Robert Bashlow, his estate, then by Manfra,Tordella & Brookes who sold it to me. I later sold the coin to Larry Briggs, who sold it to an unidentified collector. The coin has been authenticated by Don Taxay and Walter Breen. Breen examined the coin again in the 1980's, and again said it was a genuine issue. Adolf Weiss also saw the coin, and commented that he knew of another one, although the existence of another specimen has never been verified.

The reason for this coin's issuance is unknown. In any case, it is the rarest Jefferson Nickel!

The 1942S Silver Nickel with normal mintmark

1943/2 P

The 1943/2 P overdate is actually a doubled die, created when a die was first impressed with a 1942 hub, and subsequently by a 1943 hub.

This variety was discovered by Delma Romines in 1947, who was unable to get the variety recognized until numismatist Thomas DeLorey examined Romine's heavily circulated specimens at Coin Word in 1977. DeLorey was convinced that a genuine overdate existed, published pictures of the variety and asked that a high grade piece be discovered to verify the hypothesis. I was able to discover an uncirculated coin shortly after Coin World ran DeLorey's article. The uncirculated specimen allowed DeLorey and the American Numismatic Association to confirm that Romine's discovery was in fact a genuine overdate.

Curiously, circulated examples had previously been sold as overdates at public auction sales in 1972, 1974 and 1976, fetching as much as $1500.00

The 1943/2 P nickel was apparently produced as long as the die held out, since many specimens show die erosion in the fields and around the lettering. Circulated specimens are not rare, and a multiple-roll lot was sold some years ago. The coin is rare in uncirculated, there probably being not much more than a hundred coins discovered so far. Many of these are five full step coins, and one six step coin has come to light. There is a variety with a spur like diebreak on the end of the 3, at first glance appearing to be the diagonal of a 2. This diebreak variety was reported in Spadone's *Variety and Oddity Guide in 1964.*

Genuine 1943/2 P nickel

1943P nickel with spur diebreak in 3 – this IS NOT the overdate.

1943P "Double Eye"

This is a double die obverse, with the prominent feature being part of another eye showing below the portrait's normal eye.

Somewhat scarce in uncirculated, though many may remain to be discovered in unsearched rolls. The five step frequency of appearance is 1:20, with six step coins being unverified.

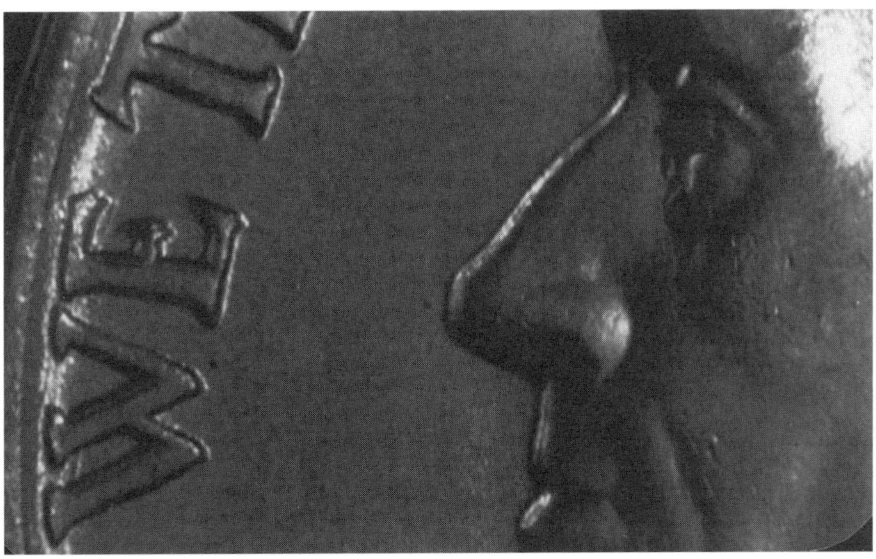

1943P Double Eye

1945P Double Die Reverse

There are several varieties of double die reverse for this year. The earliest one discovered, by the late Walter Breen in 1952, is listed in "The Red Book" and shows the most prominent doubling on ELLO and CENTS. This variety is the one normally collected. Rare in uncirculated, and not often found with five full steps having a frequency of appearance of 1:20. Six step coins may exist but are unverified.

1945P Double Die Reverse

1946 D over inverted D

This variety was discovered by Adolf Weiss and popularized by the PAK Full Step Nickel Club and has remained popular since. The variety is at this point rare in uncirculated condition, however since the 1946D nickel exists in large quantities more specimens may come to light. The frequency of appearance of five full step coins follows the normal 46D, that is 1:20. Six step coins are unverified.

The variety is sometimes referred to as a D over upside down D or as a D over lazy D.

The 1946D/Inverted D. Most show extensive die wear.
The variety is sometimes called a "D over Upside Down D" or a "D over Lazy D."

1949D/S

Produced by over-punching the S mintmark with a D, the variety was discovered about 1972. Small quantities of BU coins have been discovered, and these show varying degrees of die wear implying that the die was used for its entire life.

All genuine 49D/S have three small tool marks extending from the second corner of the balcony of Monticello. The earliest die states have a very prominent tool mark extending from the mintmark and leaning to the right.

This variety can be found with five steps, having a frequency of appearance of 1:20. Six step coins possibly exist but are unverified.

The 1949D/S Note tool marks on top right of balcony.
The tool mark extending from the mintmark is not present on later die stages.

1954 S/D

One die is known for this variety , which was apparently used until it wore out as later die states show a less prominent D under the S. One of the more common Jefferson Nickel varieties.

This variety does not seem to exist as a full strike, and few have even partial steps. The variety is unknown as either a five or six full step coin.

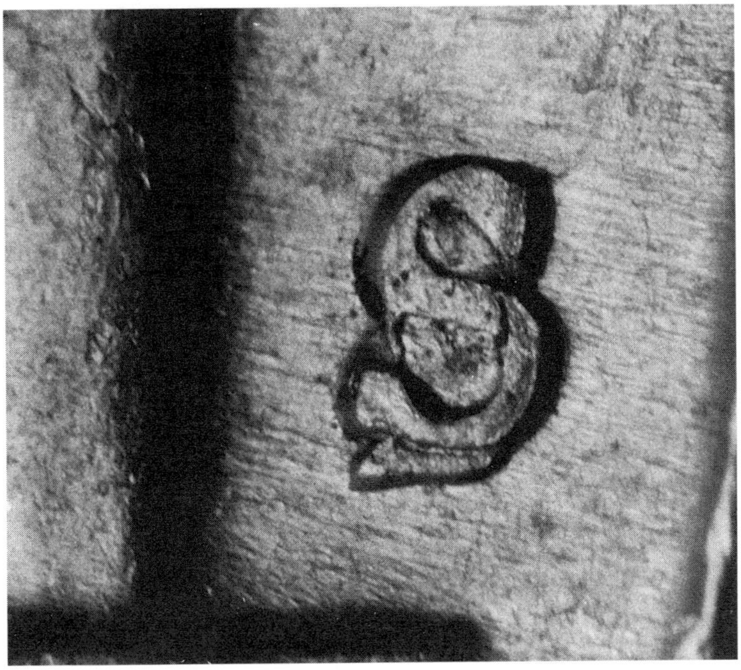

The 1954 S/D. This is an early die state.

1955 D/S

No San Francisco mint coins were produced in 1955, however apparently there were some S mint reverse dies that the Mint did not wish to scrap. This can be the only logical explanation for the existence of so many different varieties of this over-mintmark. At least ten different dies have been identified, however only the one with the most prominent S/D commands a good premium.

The 1955D/S is somewhat better struck than the 1954S/D, however no five or six step coins have been discovered.

The collectible Variety I is pictured, as well as two of the lesser varieties for comparison.

1955D/S Variety I   This is the collectible variety. A later die state is shown.

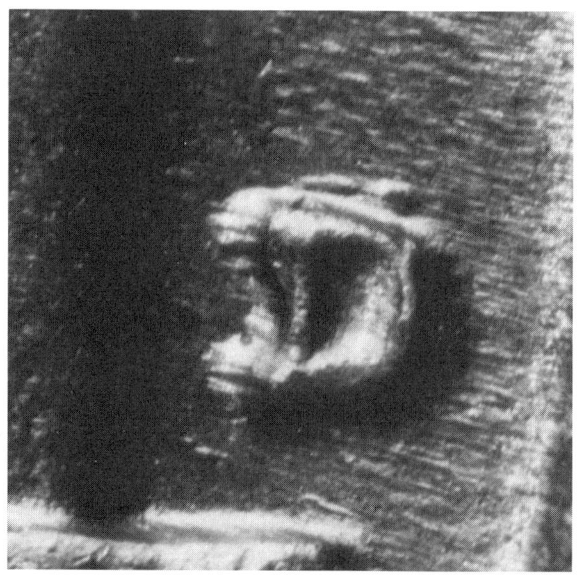

Two of the nine or so other die varieties of the 1955D/S. These have less premium.

NOTES

NOTES